*R*est *I*n *P*rint

R.I.P.

Rest In Print

MITCHELL FILBY

Published in 2014 in Australia by First Rock Media
First Rock Media
1A Lister Avenue
Seaforth NSW 2092

mitchell.filby@first-rock.com
www.mitchellfilby.com
mitchell@mitchellfilby.com

Book Production: OpenBook Creative
Book Cover Design: OpenBook Creative
Editor: The Copy Collective
Cover illustration: Based on an image from Picsfive/Shutterstock.com

National Library of Australia Cataloguing-in-Publication entry (pbk)

Author: Filby, Mitchell, author.

Title: Rest In Print / From office printing to the rise
of managed services/ Mitchell Filby.

ISBN: 9780992364502 (paperback)

Subjects: Printing industry–Management.
Printing industry. Printing–Marketing.

Dewey Number: 685.2

Printed by Lightning Source

The paper used to produce this book is a natural, recyclable product made from wood grown in sustainable plantation forests. The manufacturing processes conform to the environmental regulations in the country of origin.

To my loving and devoted family, especially my beautiful and selfless wife Wanita, who continues to maintain perfect balance and poise in all family situations.
To my best mate and son Kalahni Jai who demonstrates endless passion, maturity and insight beyond his ten years and who drives me to be a better man and father.
Also to my enchanting and stunning six-year-old daughter Jamaya Chantel, who provides me with the imagination, creativity and affection to inspire my dreams.

CONTENTS

Introduction   1

SECTION 1: FROM THE 1990S TO NOW

Chapter 1: Who remembers the 1990s?   7

Chapter 2: The millennium and a technology revolution   19

Chapter 3: The new world and doing more with less   27

SECTION 2: OPENING UP THE KIMONO

Chapter 4: The battle over the connected world   35

Chapter 5: Is it an MFD or MFP? What's the difference?   43

Chapter 6: The untold story   51

Chapter 7: A financing revolution   57

SECTION 3: THE TRANSITION FROM PRODUCT-LED TO SERVICES-LED

Chapter 8: The fight against commoditisation   71

Chapter 9: When the click rate price gives you a seat at the table   75

Chapter 10: An industry under pressure   83

Chapter 11: Managed print services just became legitimate   95

Chapter 12: To print or not to print? that is the question   107

SECTION 4: TONER IS THE REAL BATTLEFIELD

Chapter 13: More than the box   121

Chapter 14: Managed print services are not a silver bullet   133

Chapter 15: Taking the print out of managed (print) services   145

Chapter 16: Managed print service definitions and MPS 2.0   151

SECTION 5: THE RISE OF MANAGED SERVICES

Chapter 17: New managed services – red or blue ocean?   159

Chapter 18: The rise of managed services   171

Chapter 19: Services, services and more services   191

SECTION 6: WHY THE CUSTOMER NEEDS TO AUDIT

Chapter 20: Looking under the hood with a
collection of cases studies   201

Chapter 21: Five key factors to demystify a page per print contract 215

Notes   225

Acknowledgments   229

About Mitchell Filby   233

First Rock Consulting   235

INTRODUCTION

For all the readers of this book
"Absorb what is useful, discard what is not, add what is uniquely your own." — Bruce Lee

Firstly thank you for picking up this book, turning the covers and fanning through the pages. Some may be slightly intrigued and may consider taking a little more time to see if what's written is worth exploring further.

Maybe there are some revelations – gold nuggets that stir further interest, puzzlement or even potentially some excitement. I know there are some quick wins that will save end customers a lot of time and money if they read this book in more detail. As Bruce Lee suggests, if you can take something from this book, then that's a great outcome in itself.

I know very well that this book was not written to inspire world peace and I know many, including my wife, would not see this book as an interesting or exciting read. Well that is okay, it is not meant for everyone.

This book is certainly written with the business world in mind. It's written for and favours executives who need to take stock of office printing and document imaging for medium and large businesses in either the corporate world or government entities in Australia.

The wider all-encompassing term of office printing and document imaging includes and bridges a more modern scope of today's business environment. It is designed to capture the entire ecosystem, processes and ongoing management that exists between the physical printed page output and the electronic or digital document content that is created, transported, shared, extracted, repurposed and managed across multiple rich formats, anytime and anywhere.

Rest In Print: from Office Printing to the Rise of Managed Services is written and shaped for both sides of the ledger – both the end customer, and anyone who currently operates in, around or is considering entering the world of the office printing and the document imaging industry.

This book is firstly written and prescribed for senior executives in IT, such as the chief information officer (CIO); in finance roles, such as chief financial officer (CFO) or managing procurement as the chief procurement officer (CPO). Many elements of this book are designed to steer you in the right direction, save you time, money and reduce your risk and pain as you endure the multifaceted processes that exists today. In fact – as many customers continue to say – it's an industry that can be challenging, complex and confusing all at the same time.

If you are from within or around the office printing and document imaging industry you now have a documented transcript with a high level history of what has occurred over the last 20 plus years in Australia. You also now have a guide or compass to see what is currently occurring in the industry, including the challenges and imperatives that the industry is facing and what the industry has to overcome when adapting to a customer's continually changing environment.

This is the first book ever written in Australia to capture a 20 plus year timeline of the industry. It is also the first book, written in Australia on the office printing and document imaging industry, that is crafted for both the end customer as well as the industry itself.

Importantly, even though this book was written for the Australian market, it reflects the industry and customers world-wide – a true global playbook. In fact Australia today is regarded as one of the more mature markets when it comes to the office printing and document imaging industry in areas such as managed print services (MPS).

Therefore, I expect the information and insights in this book to be relevant to the global industry. As a sub-committee member of the Managed Print Services Association (MPSA), the independent peak body representing the industry across the world, I know too well that the same global challenges and future issues face everyone regardless of where they operate.

I have attempted to take a balanced and unbiased view of the industry – an industry that I greatly respect and in some weird way, have a strong affection for. In doing so I feel and hope this book provides a must-read to end customers who buy, or manage, the facilitation of office printing or are trying to move from the world of physical printed pages to one of digital documents, content and processes.

Both my passion and my commitment to the office printing and document imaging industry in Australia have driven me to write this book and capture the progress and evolution of the industry and market.

One reason for writing the book is to provide support for the industry as it transitions to a managed services culture. I also wanted to ensure the industry continues to maintain a high level of accountability and progress as it evolves and transforms.

The book itself is designed as a platform, a framework, for other experts or interested parties to leverage from. Hopefully I have provided them with a starting point. However as a framework this book extends out to many subject areas and draws on many discussion themes. I am cognisant of the fact that there is still much that I have not covered into this volume. But like all good things I hope that I have laid down a layer of fabric that will continue the story, just like our industry, which will continue to build on its never-ending story.

Section 1:

FROM THE 1990s TO NOW

CHAPTER 1:
WHO REMEMBERS THE 1990s?

SIGNIFICANT TURNING POINTS OF THE 1990s

In writing this book, I recognised that it was important to illustrate how the office printing and document imaging industry has continually changed and adapted over time – specifically the last quarter of a century. Most changes have come through technological advancement. However, more recently, we have seen the industry adopt new business models, both locally and globally, to meet the needs of new markets.

The purpose of explaining some of these critical stages and milestones is to provide a better understanding of how the industry has evolved and how resilient the industry has been to change. The intention is not bore you, but to highlight a few key points on the timeline of technological progress starting from around 1990. These are significant and compelling events that impacted and changed the Australian marketplace.

As we venture along the timeline you will see how sales organisations started to become more "creative" and mature in their business models, service support capabilities and pricing models. Some points along the timeline may explain why customers began to see definite inconsistencies in the marketplace when dealing with the industry players and actors involved.

PRE-1990: WHERE IT ALL BEGAN

It would be remiss of me not to mention two important product platforms that today compete for market share all around the world.

First, in 1949, Chester Carlson founder of the Xerox Corporation invented and marketed the first xerographic copier called the Model A.

Second, in the mid-1980s, Hewlett Packard commercialised the first single-function printers for the mass market: the laser printers that we see today.

Between and since these milestones, great companies such as Remington-Rand, Xerox (Palo Alto Research Centre), Apple, IBM and others have driven important developments.

What set the players apart was their functionality. The single-function office printer was designed to produce a set of originals (or prints) for the desktop user. These originals were then replicated on photocopiers, which produced millions of copies of printed pages worldwide.

Early (pre 1990) Hewlett Packard (HP) products like the Thinkjet were created to compete against the existing dot-matrix printers of the day. The Thinkjet was a thermal inkjet device (the first miniaturising inkjet technology) and the HP Deskjet was the first mass-marketed inkjet printer.[1] The LaserJet Printer (the first desktop laser based printing engine) celebrated its 200 millionth shipment in November 2013.[2]

Without delving deeper into history, I hope I have demonstrated the important fact that the devices that you see today came from two different worlds. One grew out of the copier world and the other, the printer world. Both are successful in their own worlds and have competed against each other over the last 20 years. Today these worlds have collided and the players will become extinct unless they can adapt to their new world.

1990: THE HUMBLE BLACK AND WHITE PHOTOCOPIER

The photocopier in 1990 was king of the office equipment world in Australia. Virtually all businesses had one. In fact, I remember my first sales manager in the copier industry would say, in every sales meeting and sales demonstration training: "All businesses need a phone and a photocopier, therefore every customer is an opportunity." With these words ringing in our ears, we hit the streets canvassing door-to-door to sell the latest new model.

At that time the demonstration was the actor's studio. The sale was all about the "demo". If you had flair and the right attitude you could combine features, advantages and benefits with some technical knowledge and a little bit of humour to do pretty well. Once the client decided to purchase the photocopier, the next challenge was to explain how the service and maintenance contract worked.

In the Australian marketplace the sale of new copiers included a service and maintenance contract. Second-hand copier contracts operated under a "do and charge" service model, where the customer only paid when the machine broke down. This was usually the most expensive option. Alternatively, the "copy kit" was a pre-paid service commitment based on a copy volume that reflected the monthly or quarterly usage by the customer.

Explaining the service and maintenance contract to a customer was not a simple process. The industry was starting to show its true colours or, perhaps more accurately, its many shades of grey.

In a perfect world, we would sell the most suitable photocopier device – based on speed and functionality – to the customer. The copy kit would be sold on a set copy volume and when the volume was achieved the customer would pay for another copy kit. Imagine having a car where you pay to drive it 10,000km, at which point it stops. You pay for another 10,000km, the supplier services the car and it will let you drive for another 10,000km before it stops again. That's like a copy kit.

The copy kit was the embryo that grew into the "contracted page" that we will cover in later chapters. Customers always bought a copy kit with their first copier. It included a set number of pages. After that ran out, they had the choice of buying a new copy kit that included all the servicing they needed for a set number of pages and may have included toner, or buying toner, servicing and maintenance as they needed it, known as uncontracted pages. A good example of uncontracted pages would be your printer at home – you just buy consumables as you need them.

COMMISSION-LED "SHOW ME THE MONEY"

The contracted pages, or copy kit, was a turning point for the industry in Australia. The copy kit – essentially the customer contract commitment – drove guaranteed revenues and profit for the sales business every month. The sales company started to generate consistent earnings. They were building their pot of gold and everyone was trying to get into the act. Most companies re-invested their earnings so they could better support their customer and build a better business.

While this was occurring the salesperson employed by the

sales companies started to see how much they could earn selling the hardware and the service. The service and maintenance contracts helped to spur on the commission plans at the time. Now depending on where you sat in the sales process this was either a good thing, or not so good. It was certainly good if you were the salesperson and sales organisation. If you were the customer, you may not have always found it the most satisfying purchase.

The sales process became motivated by sales commissions and sales revenues – probably more than the industry would have preferred. The original equipment manufacturers – mainly the major Japanese brands – did not always operate a direct sales model. It was about this time that the large manufacturers were deciding to sell direct to the market.

Before and during the early 1990s the marketplace was being developed by small business entrepreneurs who tried to generate revenue the best way they could. Their low-cost base and higher incentives were a realistic way to build scale and this model still works for small and large businesses today.

However, because sales commissions are a proportion of the sale, the salespeople were motivated to sell ever larger and more expensive copiers and copy kit volumes to maximise their income. Customers were being conditioned to buy larger and more expensive devices. Because larger photocopiers were more powerful – to produce large copy volumes when you needed them – they were marketed as more reliable. At the same time copy kits on large devices were better value than smaller devices, which carried a lower copy kit threshold or toner-yield per bottle. This cost benefit model is still very much the norm today.

Another layer of complexity in the sales process was the number of variations to the copy kits that were available – surprise, surprise. Here are some examples:

- A copy kit was for a set copy volume or a maximum of 12 months (whichever came first).

11

- A copy kit could include all parts and onsite service.
- A copy kit could include some specific parts (or no parts) and additional parts were chargeable items outside the copy kit.
- A copy kit could include toner, or toner could be purchased separately.
- A copy kit cost could increase in price based on any of the following:
  - the age of the device measured over time
  - the age of the device measured in copy volume
  - producing copies over the set agreed volume for the copy kit, and/or
  - scheduled annual increases.

Many copier companies thought they were entitled to charge whatever they could for toner. You may call it creative marketing (they are in business to make a profit, after all) but the industry pricing model became inconsistent across brands. Customers could not understand the basis for charging and began to distrust the industry as a whole.

In that scenario, everyone loses.

The copy kit pricing model started to lose favour to the contracted pages model – where you paid a cost per copy or price per print. Variations on these contracted pages models are what the industry uses today.

THE COST PER COPY PRICING MODEL EVOLVED

The industry in Australia embraced the cost per copy (CPC) model in the very early 90s. The cost per copy pricing model provided the customer an exact price per page for every page they copied.

The page per print (PPP) pricing model you see more commonly today was the advent of the digital revolution of the mid-90s where copies produced moved to pages printed; "print" being the definitive word.

It also emphasised how the industry globally adopted a new more modern terminology, especially with the battle between the photocopier-based manufacturers and the single-function printer manufacturers. This battle is explored as a separate chapter later in this book.

THE COLOUR PHOTOCOPIER

Although it was produced a decade earlier, the colour photocopier didn't pick up speed until the early 90s in Australia, largely due to consumer perception of poor colour reproduction and high cost of consumables. Also sales of black and white photocopiers were booming. It is hard to recall exactly when the colour copier finally achieved traction, though the leader in colour copying in Australia at the time was certainly Canon. However, the battle for colour was just around the corner as the industry took one giant step into the digital world. Digital colour would see both Canon Australia and Fuji Xerox Australia dominate the market over the next decade or more. When black and white office production devices were added to the mix (typically 60 pages per minute and above was considered production) the battle for office domination was raging.

THE DIGITAL WORLD AND THE BIRTH OF THE MULTIFUNCTIONAL DEVICE

If I had to choose a time where the worlds collided for the traditional photocopier salespeople and industry in Australia, I would choose 1995. At that time Fuji Xerox officially launched the first digital (networked) copier. It was sold as a photocopier replacement. We were all a bit unsure of this device. It looked different, it apparently could copy and print at the same time. "But how could it do both at the same time?" we thought. I think this is why it was only marketed to the sales teams as a digital copier. Maybe the marketing teams could not get their heads around this new technology?

However, what seemed difficult at first became easier and easier to position with clients. Other brands started to market their own version of this digital copier – later known as the multifunctional device (MFD) – that could copy, print, fax and scan concurrently. MFDs were designed to have the same functionality as the client's previous photocopier. It could carry both A4 and A3 sheets, as opposed to most printers that only printed on and from A4 sheets. MFDs were only sold by the traditional photocopier manufacturers. The copier industry as a whole was very resistant to calling it a printer because printers came from the world of its future competitor.

Because it was connected to the network, the MFD brought a new perspective – and a new competitor – into view.

First to arrive was The Docutech: large, commercial Fuji Xerox Production units that changed the commercial printing industry, dominated office print rooms and allowed instant printers to provide a faster service to customers.

Second, colour moved into the mainstream with Fuji Xerox and Canon battling for market domination through the end of the 1990s and into the new decade.

Third came the realisation that the industry was actually bigger than we thought. We had a new competitor that owned the IT world. Hewlett Packard (HP) was the product and brand of choice for network printing. We were told IT managers only put HP printers on the network. Sure there were a few other brands like Lexmark and Epsom but it was dominated by HP. As we opened our eyes wider and refocussed our lens, we realised that printers were everywhere – they bred like rabbits and still do in many organisations. These devices lived in the world of uncontracted pages.

These devices – no, these printers – were bought through IT budgets; not the office budgets our clients used to buy photocopiers or the new networked multifunctional devices. IT managers were making the decisions and not the office manager or procurement, who were our traditional point of contact. We had a new customer; one that we did not know and whose language we didn't speak. To us it was a foreign world.

A new battle now would rage. We had a new competitor and a new buyer and this would force us to adapt and do it fast. We had to be smarter. We had to muscle up and it was in our competitive nature to do so. It may be insignificant to many who read this, but for the passionate people who worked within the industry this was their life, their profession. They were the most well-trained sales force in one of the most competitive industries. I don't think any industry invested in sales training as heavily as the copier sales industry between 1986 and 2000 in Australia.

I recall being the general manager of sales and marketing for a publicly listed IT company in Australia and interviewing a number of heads of IT departments. These senior people were called chief information officers (CIOs). Most of the CIOs mentioned that they received constant calls and canvassing from sales organisations. When I asked what industry the sales approaches were coming from I expected an answer like Cisco, IBM or the like. What I didn't expect was the reply I received.

A CIO from Deutsche Asset Management, who was more outspoken, said he was constantly being contacted by "those copier guys, they never stop. If there's a sniff of a deal they are all over you. It's like you tell one salesperson and he goes out and tells all his friends and then you have all the brands at your door, 100 deep."

At this time I was shaking my head to emphasise that I understood his pain, frustration and annoyance. I was certainly not going to volunteer my past history and training at that time. But then he stopped. I waited a moment and then said to him, "How often would you receive calls from your other IT suppliers?"

He then said something I have never forgotten. He said, "I never hear from those guys ever. I'd be lucky to hear from them once a year. Even when I want them I can't get them. It is so hard to buy from them because they never come back to me."

"Wow," I thought, "that would never happen where I came from. What a great opportunity this IT industry is."

He then went on to say, "You know it might be bloody painful at times to have these copier guys constantly calling. But I'd rather do business with people and an industry that has a passion to sell and wants to sell me something."

That rocked my boat. As an industry with all those perceived concerns, we were punching above our weight. The 1990s built a foundation and resilience into our industry and people. The challenges that we faced in the 1990s were only preparing us for future battles when the playing fields were not so clear and known.

Later towards the cusp of the decade we would see the launch of the "mopier" from HP. Mopier stands for Multiple Original Prints (the "ier" is a throwback to the analogue "copIER" term). But analogue it was not – it was certainly a digital technology. I guess this product launch name was designed as a strategy and messaging play for the printer companies such as HP to fight back by trying to win the front office copying and printing market. The mopier was targeted at larger workgroup areas within the office. It would lever its performance as a desktop printer and include scan, copy, fax and email service functions to become a more integrated device.

In short we may have disturbed the beast. The printer suppliers were now actively pursuing the office volumes where the traditional copier (now MFDs) had dominated. This was the start of the MFD and multifunctional printer (MFP) battle for supremacy. The "P" in MFP actually stands for Peripheral. Personally I prefer to use the P for printer to recognise the device's heritage.

CHAPTER 2:
THE MILLENNIUM AND A TECHNOLOGY REVOLUTION

2000: THE OFFICE COPYING AND PRINTING DEVICE UNDER THE MICROSCOPE

The world would quickly move on from fear of the Year 2000 Millennium Bug. IT would solve all our problems. It was unstoppable. It was our future.

We thought...

The world changed for all of us on September 11, 2001, especially in the United States.

Until then growth had been the focus but as the world recoiled, so too did businesses built on the theme of "Build it and they will come". The cracks soon appeared and financial rigor became more the norm. Organisations started to buckle under reducing cash flows and debt. Shareholders were being exposed to investment risk and reality. Liquidity was drying up. In the end many businesses, big and small, went out of business. Increased financial transparency measures started to expose how badly some businesses were run, why they went bust, and how success or failure were sometimes down to timing, rather than actions.

Through an increasing focus on financial management we started to see a number of organisational restructures and business transformations. Although the chief executive officer provided the vision and drove the strategy of a business, we increasingly saw chief financial officers drive change for financial reasons, business survival and shareholder approval. This made sense as, behind the C-level executives of these large enterprises, shareholders were watching every move, every profit and loss result and the strength of the balance sheet.

Through this attention we started to see changes in the chief information officer position; shifting from IT-driven technologists to more strategic and business-orientated people who were comfortable with and understood IT. No more triple redundancy. "No technology for technology's sake"; "Technology should be the enabler for the business"; "Do more with less", were all catchcries heard when the rhetoric changed.

As the business changed its reporting structures to provide improved transparency of spend across the business, we saw IT, Procurement and Finance come together. In some enterprise organisations we saw strategic sourcing or shared services teams formed to provide a central point of control, management, synergy and cost rationalisation between the three business functions. Although many of these teams still exist, and some have been created more recently, a high number did not last due to lack of cultural team integration, sponsorship alignment or business unit support.

I remember in 2004 meeting with one of Australia's largest finance institutions (one of the four pillar players) and discussing how they were going with their shared services unit. This financial services firm had a clear mandate on how it should function and what the expectations were. In fact, this business had 16 different business units vying for the business's financial resources. Each business unit had a general manager and each had the same objective: profitability.

This created a challenge for the shared services business unit because each of the business units could run autonomously to achieve its outcome. One of the main drivers for a shared service model is to gain cost advantage and efficiencies by acquiring technology or services to be used across the business. Put simply, the buying power becomes the advantage for the shared services model.

When cost efficiencies become one of the overall business drivers, it makes it more difficult to build unique technology products and services for an individual business unit. The costs, time and resource requirements impact the other business units that are not involved. This makes the individual business opportunity less attractive for the shared service business unit to undertake the project. When this occurs, a shared services model risks a swift collapse if individual business units continually omit themselves from the process, or feel that shared services does not provide them the innovation, agility or capability to compete more effectively.

In the case of this financial institution, shared services could not be sustained in its original form because too many business units had opted out and executive sponsorship had become more about the profitability than about the business unit itself. In the eyes of the executives, if shared services got in the way of making a profit then it was not a workable outcome.

The key take away from this point is that salespeople need to understand the business and operational drivers that exist and appreciate that the business can be very diverse. At the same time if an end customer cannot effectively bring all internal areas together (i.e. centralise buying decisions) they will miss out on the real value of leverage.

THE DEVICE IS NOW AN IT ASSET UNDER THE MICROSCOPE

The increased financial rigour put the MFD and printer under the microscope in terms of who buys the technology within the business, how is it purchased, how many devices the business has, who produces the printed page and why, and how much the business prints in black and white and colour. Also, who pays for the running costs and who manages the device once it's installed and connected to the customer's IT network. These questions may seem obvious to many, but I have seen many customers who are totally unaware of what, who and how costs are being charged and managed.

These questions and answers were not exclusively driven out of this new financial stewardship, nor were they the domain of enterprise organisations alone. Organisations were reviewing and developing best practices in a number of operational disciplines. Answers became clear in organisations that were aligning and combining IT Services, procurement and finance. The group was more likely to reduce cost and increase efficiency by focusing on operational and process redundancies, cost duplication, vender consolidation, IT device standardisation and model rationalisation – just to single out a few areas of focus.

With this focus now on the device asset the obvious started to appear: nobody was solely responsible, costs were not being managed, there were pricing inconsistencies and benchmarking of a contract across its contract life was non-existent. Organisations had no way to control, forecast and budget for their internal printing environment after year one. It was a moving target. The term "if you can measure it, you can manage it" now started to apply to the print environment for the first time. Businesses needed to ensure greater predictability in their operational costs and printing was too variable.

AN INDUSTRY SLOWING

By mid-decade the office printing and document imaging industry in Australia, dominated by the larger tier one brands at the time (Fuji Xerox, Canon, Ricoh, HP), was showing signs of slowing revenue growth and business momentum. Most of these larger tier one brands had taken considerable action to reduce costs from their own internal business operations. They improved and integrated more effective enterprise-level software and systems across their business which, at times, nearly killed the business in mid-stride.

Over this period a number of product manufacturers entered the arena of enterprise bargaining with unions who represented the field service technicians. By negotiating employee contracts and entitlements, the brands aimed to provide more flexible service support for their customer. They aimed to reduce their internal costs by increasing the service technician's daily call productivity and providing incentives to be more efficient in-field in areas like service response commitments, number of calls or jobs per day, first-time fix, parts allocation and replacement and call management.

The backdrop for these actions was, and continues to be, the high costs to manufacturers and large dealerships of supporting direct sales, marketing and service operations in Australia.

An outcome of the slowing and more competitive market during this period was manufacturers, dealers and service providers competing more aggressively on price. They changed their internal structures to grow and encourage sales activity and market share.

For example, we saw over a few years the colour service click/ print rates come down from about $0.24 cents per print in the early part of the decade to about $0.10 cents per print towards the end of the decade.

As competition for business heated up we also saw a number of large Australian corporate enterprises such as Telstra, Westpac, CBA, IAG and AMEX adopt their own version, or a supplier's version, of what was known as fleet management. Fleet management was usually a single-source branded solution based around hardware, a small amount of software and professional services and a service contract. Fleet management was the precursor to the more sophisticated MPS that exist today.

This period was extremely important for both the industry and the end customer. Over this period we saw customers recognise and adopt the fundamental elements of a more encompassing service delivery model. They also started to see that the multi-functional devices were about more than print and copy; they could be an intelligent information hub for potential business processes and services.

In 2003 I was asked to assemble a combined team from Canon Australia (sales) and Canon Information Systems Research Australia (CISRA), to develop the concept of the "Information Hub". I handpicked an external consultant who was a highly successful and entrepreneurial thinker to lead The National Enterprise Account team. With the help of CISRA's lead commercial manager, we developed the product over 12 months, then engaged and demonstrated our new technology to 25 chief information officers.

Their feedback was tremendous and Canon Australia won accounts with Telstra and Westpac as a result. They were two of the largest and most sought after contracts to go to market. Today I guess it's the norm. But it does show it takes time for the stars to align and acceptance to mature, but in the end it has to start somewhere.

All this made me realise one thing – one "Ah ha!" moment. This technology and what we could provide for our customers was no longer about the device and its functions.

This technology was more about what sat in the device; its

software or firmware platform. The platform allowed the device to be customised to the customer's workflow. It delivered new client-enriched applications or services far beyond the simple functions that everyone could provide. Business services and applications could be customised for an individual user. Processes could be automated or streamlined to increase efficiency, improve process transaction speed, enhance customer satisfaction, reduce errors and allow quick and more reliable access to source information through centralising information and data. The list goes on.

A new beginning...

CHAPTER 3:
THE NEW WORLD AND DOING MORE WITH LESS

2009: GLOBAL FINANCIAL CRISIS AND THE CONCEPTION OF MANAGED PRINT SERVICES

By 2009 the world had been hit by the global financial crisis where many economies and industries around the world were starting to feel the massive impact of structural change. In Australia this was a blessing in disguise for the office printing and document imaging industry.

In that same year a small United States analyst firm started to show its head and, through its endeavours, put together the first MPS conference. Although there had been many differing terms and acronyms within the industry, this was the first time an independent organisation had formally put together an MPS conference. Out of this conference the MPSA was formed and developed a definition of "managed print services". (See Chapter 16 for the definition.)

THE BIRTH OF MANAGED PRINT SERVICES

The industry saw MPS as a new vehicle to leverage their sales and marketing activity, which had slowed in the previous few years.

MPS would become the next growth engine in Australia, mainly because managed print services provided greater transparency and accountability of the office print environment. As such, end clients could now better understand what their real costs of office printing were. The visibility of their devices and their associated print volumes could then be matched against the actual costs of office printing. This ability to measure and manage your internal office printing allowed a number of benefits to the customer, the main one being cost reductions. MPS is certainly a strong vehicle for the customer to reduce their costs. Many managed print service providers continue to market and quote 30 per cent savings, but I expect that this percentage is lower for customers who have already been through their first MPS contract. However, for customers who have not seriously looked at an MPS solution, I do expect cost savings to be significant.

From 2009 the cost savings mantra became the value proposition for the global industry. The manufacturers' sales machines went into overdrive. MPS matched the client's objectives perfectly at the time. The global financial crisis meant that costs had to be taken out of the business quickly and easily. MPS was that vehicle: a win-win for both sides of the ledger.

This chapter is focused on the timing of industry change and I will provide more substance to MPS in later chapters. For now MPS is the start of the services-led business model that many device manufacturers, dealers, IT service providers, office supplies firms and the like are trying to navigate.

CHANGE OF GEARS?

From about 2012 onwards, the global financial crisis came off its most critical watch period in terms of both legal and financial proceedings and people began to accept the new status quo. We observed a new level of organic growth, reinvestment, acquisition, mergers and repositioning across the office printing and document imaging industry. In addition we saw heightened activity from external industries that impacted our industry and markets, as they made plays into our industry and market heartland. We also saw ongoing investment activity from within the industry in Australia.

Looking towards Europe, we saw a small number of European countries still working with high debt levels which still play havoc with the global investment market today. This may be continuing to damp down investment and acquisition levels. Looking towards the US, we see the beginnings of a rebound in the activity cycle and the once critical economic stimulus measures are starting to taper off as jobs increase. This will boost confidence and I expect some major moves by a number of leading industry players as a result.

A good indication of improving sentiment in the US is the feedback that I receive through my fortnightly MPSA education committee conference calls. US Members of our committee are regularly reporting that employment rates in our industry are increasing at a rapid rate. I have a couple of theories about why this is happening but they would fill another book.

The MPSA is an independent, not-for-profit, organisation that represents the global printing and document imaging industry. Members work hard to enrich and educate the industry and end customers – like what this book is trying to do for Australia. If you are interested you can find out more at www.yourmpsa.org.

THOSE WHO DIDN'T COME

Possibly more interesting than what did happen is what we thought would happen, but didn't. From 2012 we expected IT managed services firms across the world to make a bigger and more strategic play to move into the office printing and document imaging industry, to strengthen both IT infrastructure and IT services businesses. The attractiveness of increased scope generating new revenue, annuity streams and widening their service delivery capabilities looked like a logical deployment, both to defend their business models and as a strategic development.

It was a natural play from the sense that the customer saw this as a natural progression too: "You already operate and manage my infrastructure, you may as well run the devices and printers too. Just wrap this under our existing services agreement and you can manage it from there."

Well it didn't happen to the degree we thought it would.

In fact we have seen more of the reverse (you can't keep a good guy down!). In reality the manufacturers have taken an early-mover advantage and invested by purchasing existing IT services firms or they built the capability in-house, ready to create a more services-enabled operation.

One of reasons why some IT managed services firms did not take this strategic route is because they chose to move in a direction they saw as more exciting and more aligned to their vision and passion. They choose to move further upstream into the world of cloud services, business intelligence, and security. These areas were building enormous excitement, energy and momentum. In short, print was not sexy; print is just attractive to the IT purists.

2014 and onwards brings both more challenge and opportunity. A number of manufacturers in Australia will face the challenge of reducing their operating costs for their direct business, while others will continue to build and differentiate their programs that support dealer, agent, reseller and distribution channels in a bid to grow market share.

Section 2:

OPENING UP THE
KIMONO

CHAPTER 4:
THE BATTLE OVER THE CONNECTED WORLD

THE BIRTH OF DIGITAL: THE MULTIFUNCTIONAL DEVICE IS BORN

The photocopier was king. We were the only game in town. Little did we know...

In 1995 everything changed in Australia. Our beloved photocopier, that we knew so well, came from the analogue world (it used an analogue signal/voltage/pressure and process rather than a digital signal). Analogue was a term we didn't even know existed prior to 1995. We were now entering the world of digital – a connected world – and we were late to the game.

I was at Fuji Xerox Australia and a product called the Fuji Xerox Able entered the ranks. At first we were a little unsure of this new type of product. It was digital. It was connected to the wall by a blue cable as well as a power point. That this device had to be connected to a port in the wall sounds obvious today but it didn't then. Just short of 20 years ago this was very new to a lot of people outside the IT world. We'd only recently switched from green screens to colour.

"Gee, how will we do this?" we thought as we struggled to take in the concept. As a sales team, either in sales meetings, sales training or at lunch, we would pose such questions to each other. There was lots of new technical information we had to grasp, including how the device functioned and what new benefits it offered to the client. We continued to see some very simple points raised regularly by both sales teams and customers, "What happens if there is no port where the customer wants the device?" "Who has to pay for the cabling?" and "It only comes with a three metre cable!"

I know looking back on this it sounds absurd that this was the focus. But in those days copier devices just came with a power lead so it was not too far beyond us in thinking we had to supply the blue cables as well. In fact our customers (procurement and office managers) asked us who was paying for and supplying the cables. Not sure why this kept coming up, maybe it was the opposition? Maybe it was those who were selling those laser printers to the IT department? They had us jumping all over the place for a time.

A number of salespeople couldn't get their heads around the whole digital technology thing in those early days. It was like some didn't want the change to occur; they liked things how they were. In fact, it became too hard and too high-tech for some.

The rest of us embraced this change. Not just because of the potential new sales but because we saw a way forward with technology that was smarter and potentially more reliable than what we sold in the past. The old photocopier didn't really have a fantastic name when it came down to reliability: it was very mechanical, with lots of moving parts, and lots to go wrong.

So we could see the digital world as one that would benefit both end user and their business. As a salesperson in those days we just hoped the new machines wouldn't break down as much so we wouldn't spend as much time apologising to customers and running back and forth between the customer and our service department. That was a job in itself.

Looking back it just shows how the industry was still very insular and box sales orientated at the time.

READY, WILLING AND ABLE?

Now it was strange that the first digital model range came out as a digital copier called the Able. Pretty inspiring stuff don't you think? Or maybe not. Who knows where they got that name from, but no doubt they spent a lot of marketing money and time to come up with that beauty.

At the time, Able was marketed as a copier replacement and nothing more. Later the Able range rolled out with a number of different options such as copy, scan, printing and faxing. Different model configurations followed, some with limited feature capability such as print, scan and copy (no faxing) and others with full capability on line. I have never been sure why they brought out different versions; maybe they were still trying to figure out what this product was really about. Maybe they thought this was a better way to evolve our skills, capabilities and knowledge.

However, when the Able became fully functional (with copy, print, scan and facsimile all on board) we had to learn the word concurrency (another way of saying multitasking or just doing multiple things at the same time). It could do it all and at the same time. Sounds easy but how hard was this to explain back then. You could copy and perform another function at the same time. The customers looked at you like you were a magician. They already knew we were good with smoke and mirrors, but this was taking it to another level.

The customers were all trying to figure how it worked. We

scan a document, fax it out and print something at the same time, marvellous. However, as we did these demonstrations – and, yes, the demonstrations were still a hangover of the photocopier training process of the past – we discovered that in fact customers had to see it to believe it.

Even our demonstrations had to be managed differently in the beginning as the sound of the device was much quieter and the speeds appeared to be slower as the image scanned only once regardless of how many copies or prints you required: the light source only moved once across the page, not 20 times if you were making the 20 copies from our old beloved photocopier.

So the client's first reaction was that it looked much slower. Although it wasn't, they were just so used to lots of activity as originals flipped over and around in the document feeder. They were also used to the noise of the device clicking and pushing through the unit, going around again especially if it was printing on both sides and then seeing the sheets spitting out onto the exit tray. All the time waiting for that paper jam to happen. But it didn't.

WHERE IS IT?

In another surprising twist, what we also observed and realised was that our customers didn't really change. What I mean is that the person who made the decisions on the photocopier was still making decisions on this new connected digital copier multifunctional device. No sign of any IT person in the process, yet.

The office, procurement or finance managers (regardless of whether it was corporate enterprise, government or small business) were still our go-to people for this digital copier multifunctional device. Although we would have a section on the paperwork that IT would complete (later it would move to a page or more) the sale was still managed through our regular contact. As other traditional photocopier manufacturing brands started to arrive in the Australian market with their own versions of digital multifunctional devices, the marketing buzz around this new break-through product became louder and faster. From then onwards both the customer and the sales force were becoming more mature in their technology conversations. Customer acceptance of this new digital multifunctional device was increasing at a rapid rate. It was out with the old copier, in with the new device.

As this occurred a noticeable change became more evident to us all. IT was now starting to be vocal. It started in the background but then moved to the forefront of many conversations and meetings. I only imagine that the business was connecting more and more devices to the network and IT were being over-run with issues such as help desk calls, IP address releases, port availability and so on. I know there were also questions inside the end customer organisations about who was paying for the physical device itself and who was paying for the toner that the device used.

This was forcing IT to become more engaged. Typically IT assets may have been funded by IT while the division or business unit payed for the toner. That was okay when the hardware asset, a small printer, was a couple of thousand dollars. When the cost increased to tens of thousands of dollars IT had to stand up and ask the question: who is responsible and accountable for the device when it's attached to the network? More importantly, who's paying for it and which budget is it coming from?

WHO'S BEEN EATING MY PORRIDGE?

As we started to learn more about this connected world we started to widen our radar and focus. As we did we saw an unforeseen opportunity as well as a huge threat and competitor to both our existing and future worlds. A market we really didn't know existed in the past. Well maybe that's not fully accurate, I think it was more denial: an idea that we sold big, metal, A3 devices and the other guys sold little plastic A4 printers that we saw as (almost) disposable.

In reality, sitting next to virtually every computer was a small A4 desktop printer. It produced mostly black and white A4 single sided images. These devices were supposed to print off one set of originals from an application such as Microsoft Word. Then you would use the copier to produce multiple sets of this document.

We realised that there was a lot more printing being done on these single function printers. This was the start of an ongoing battle. It was game on as every brand wanted to move that printing volume to their device. You see, the money was in the toner that was laid out onto the page. The more you print, the more toner goes onto the page and the more money the suppliers make.

Traditional copier manufacturers offering the new, larger, multi-tray options (A3, A4 and other sizes) for paper stock were not only competing against each other. They were going after the enormous volume of A4 printing produced on these single-function laser printers.

The likes of Hewlett Packard and Lexmark, the real hallmark brands of office laser printers, were now fighting for their territory. These new digital MFDs could replace two, three or more laser printers in one go. Sounds simple but never underestimate the human psyche when it comes to ownership.

I remember a number of customers telling me stories about their business and how their staff reacted when they tried to

OPENING UP THE KIMONO

take their single functional laser printers away. They said you can
change their furniture, change their PCs, change the phone system they use, but you try and change their printer – they become
uncontrollable, they go into some sort of frenzy.

It turns out their printer was a very personal thing. This device
rationalisation would be a much harder task than we thought and
it took more time for the compelling argument to win the day.
Luckily we had a thing called software that allowed the MFD to
address some of the major concerns that users had about privacy.

It would only take a few years for those printer companies to
fight back through the launch of their own multifunctional device.
The first was the Hewlett Packard Mopier that was positioned for
larger user work groups, the copier and MFD-centric customer
targets of the past. However these multifunctional devices were
not called MFDs: they were known as the multifunctional printers
or MFPs and they were going after the same customer with the
same marketing message.

Now we had more brands and players going after the same
market. Surely that wouldn't be a problem? People will always
print. Right?

The battle lines were well and truly drawn and the battle continues today without abate. The only thing different is that the
market opportunity may now be starting to shrink. What happens
to the swimmers in the pond when the water starts to dry up?
What will they players do? Will they adapt or die? Will they take
out their opposition in the process? To paraphrase Megginson,
paraphrasing Charles Darwin, "It is not the fittest that survives,
but the most adaptable."[3]

CHAPTER 5:
IS IT AN MFD OR MFP? WHAT'S THE DIFFERENCE?

WHO'S CONFUSING THE CUSTOMER?

I have been asked by hundreds of customers for an explanation of the difference between a multifunctional device (MFD) and the office desktop printer that has now morphed into a multifunctional printer (MFP). This is not me telling the customer they need help; it is the customer telling me the moment I walk in the door: they need help.

The customer has become so confused over time after meeting ex-copier salespeople (now selling MFDs) and ex-printer salespeople (now selling MFPs). Customers hear different spiels about how the devices and technology are different from their competitors; the various pricing programs, service breakdown costs and warranty levels. By the time they have spoken to half a dozen salespeople the customers are usually more confused than before they started the process.

One of the elements that leads to this confusion is less experienced sales operators leading with the service print charge rate before they understand the client's business needs and requirements. In delivering this simplistic pricing offer they open up discussions about why consumables and toner for an MFD are cheaper than for an MFP. Or they open up discussions about why an MFP is more economical because it's half the capital cost of the MFD.

Now I know that the good salespeople are not having this conversation off their own bat. Some are very much drawn into this web of conversation by the customer's confusion and are only trying to help navigate the customer through to a sensible resolution. Unfortunately this sometimes adds to the confusion and makes it worse for the customer.

It may sound like I am trying to suggest there are two types of salesperson selling two different product lines. That is not my intention, although it does happen more often than not. However, as we move to a services-led engagement I expect to see this less and less.

I have segregated these two product lines so I can better demonstrate later some of the key differences between the technology platforms. As mentioned in previous chapters the MFD is derived from the photocopy-led brands such as Xerox, Canon, Sharp and Toshiba. While the MFPs were developed out of the single function printers that came out of companies such as HP and Lexmark.

I will explain some of these differences in the hope it will help us all to move forward.

WHAT YOU SEE IS NOT WHAT YOU GET, OR MAYBE IT IS?

I guess these are not really definitions. They are more descriptions of products lines or technology platforms. However, to make sense of the office printing and document imaging industry we need to provide some level of education. Let's make a start.

To the uninitiated a digital single functional printer (MFP) and a multifunctional device (MFD) are the same technology. They kind of look the same although they could differ in physical size, they perform very similar functions in that you put paper in the paper trays, you either select or press the enter key or print key on your computer or you hit the green button on the device to print. As a result paper comes out onto the exit tray. That's all pretty straight forward.

Based on this it should be quite simple to source the brand, model and product configuration you want. You would expect to have the same pricing standards or metrics to both the purchase price and the service and maintenance contracts.

Sorry, no, they are not the same.

HIGH LEVEL DIFFERENCES

Firstly the single function printer (which can morph into the multifunctional printer) and multifunctional device (as mentioned, it morphed out of the photocopier) were designed and built originally from two different technology platforms and for two different purposes. This is where most of the complexity lies within our office printing and document imaging industry today.

Due to these fundamental differences it means both the hardware configuration and the associated capital are not consistent across both product sets. This also extends to how the different devices are supported through different service level contracts, how maintenance and toner consumables are used, supplied and charged and what is included under device warranties. Even the physical size of devices can be different, although the trend from both camps is to make the devices smaller.

This difference and the level of variability and functionality mean that the charging metrics are not necessarily comparable across both platforms.

Secondly the photocopier or MFD was designed to support larger paper sizes such as A3 paper, hold more paper stock in its paper trays, and provide increased functionality and flexibility for a greater variety and volume of document reproduction.

On the other hand, the single functional laser printer supported the increasing demands of the personal computer user as they provided fast, high quality text printing with multiple fonts. These devices were also regarded as more reliable due to their simpler paper path and paper stock standardisation. However it was intended through their original design and performance that these single functional devices would produce document originals only and not production-level volumes. (This intent changed once the printer brands saw the income streams they could make from the toner usage.)

Thirdly the MFD (from the photocopier heritage) was usually sold through both direct and indirect channels. Direct sales were through the manufacturers themselves and their selected resellers and dealers who were generally trained and certified to provide the service and maintenance support required. These photocopiers were previously more service intensive, so all man-ufacturing brands needed to provide highly responsive service and greater coverage.

As these devices generally produced more copies, the impact when they were out of action became more critical to clients. This therefore culminated in manufacturing brands, which sold the larger A3 copiers, to provide a service and maintenance contract that included some onsite service.

Non-authorised resellers and dealers in Australia can find it difficult to source genuine parts and supplies for newly released products. This has been one of the reasons why it difficult for a second hand market to exist in Australia especially for newer devices.

Due to the previous photocopier legacy, MFDs have been burdened with the same service expectations and commitments. A point that I believe may change in the future. Where else can you buy a product and have a guaranteed onsite service and maintenance (including parts) for five years as well as a response time of around four hours?

The single function printers were generally sold through an indirect distribution and sales channel model such as IT resellers. The main reason for this distribution channel is because these devices were usually a low cost, plug and play installation for many IT staff. Ship in, ship out and then let the customer install, repackage and send back.

Service requirements of single function printers, if required, could be managed by a number of alternative service suppliers. The expectation was that the printer was a much more reliable device than its photocopier counterpart and some clients never took out any service warranties aside from the standard manufacturer's warranty which usually offered only back-to-base support. In this situation the onus was on customer to return the device for repair or replacement.

It was typical for a customer to have multiple printers on site. If or when there was a breakdown, the urgency of fixing this device was usually lower because the ability to print to another device was often available.

Service warranties, or care packs as they are sometimes called, could be provided through the sales channel that sold the product. Or they could be delivered through the brand itself. In many cases the brand will provide a service capability delivered through an outsource partner for the break and fix service.

To reduce the service intensity requirement and costs of rapidly servicing so many devices in the field, the manufacturer was able to provide a number of service and maintenance level packages (care packs) that the customer could choose from, such as same day or next day service.

In the early part of 2000 we started to see increasing success of the multifunctional device taking hold in the front office environment where the photocopier traditionally had a stranglehold on the market. As this success was happening the direct sales force from the MFD manufacturers were increasing their penetration into the back office (the traditional home of the single functional printer). However this was not about device market share. It was about print volume market share. HP at that time produced more print volume in the office environment than any of its competitors. It could be realistically stated that HP had a monopoly in the back office.

As the threat increased we saw the evolution of what is now called a multifunctional printer (MFP). The single-function printer manufacturers developed their own multifunctional device based on the traditional laser printer engine. At the same time the traditional MFD manufacturers introduced the equivalent MFP competitor version.

Looking at the cost side of the equation, the A3 capable MFDs have typically been the more expensive option in capital outlay compared to their smaller MFP counterpart (potentially double the price), especially when configured with increased functionality such as bookmaking, larger finishing units, or increased stapling options.

However, due to their size, MFDs are usually more economical to run in relation to toner usage capacity and toner yield (efficiency) and therefore run at a lower service rate. If volumes are high, these healthy cost savings each and every month offset the higher initial capital cost. An indication of this is that you are often given a flat rate of service for onsite service, all the toner you can eat and all parts and labour for five years.

The smaller A4 MFPs require a lower capital outlay but have more expensive operating costs compared to their bigger brother MFD, for precisely the opposite reasons stated above. Warranty periods, parts, maintenance and service availability all differ depending on how they are packaged.

Both devices may look similar, function and perform a task or job in a similar way but the real differences lie in where they have originated from, how they were built and how they were meant to perform. They have different mechanisms or ingredients that make up the units and this has translated into cost and pricing differences for customers.

CHAPTER 6:
THE UNTOLD STORY

WHAT THE CUSTOMER NEEDS TO KNOW

It has been demonstrated on many occasions that both failure and success can breed and drive new levels of competition and innovation. In a business context it can allow organisations to experiment with new products and services and new means with which to deliver them. It can enlist new business strategies and business models to capture more of the market it hopes to dominate. This may be introduced so organisations can maintain their leadership position in their industry, or it can be utilised for a competitor to gain an advantage over an existing industry leader – even to reposition, or propel, the business into other new markets.

Unfortunately, in some cases this action does not always produce the right (or desired) outcome and what it leaves behind is a stain in the hearts and minds of the customers and the actors who operate in that industry.

The office printing and document imaging industry, both in Australia and New Zealand, weathered storms of distaste, tarnished reputation and customer distrust from the late 1980s through to the early and mid-1990s. Fortunately strong business practices by the leading industry players have checked much of the unethical behaviour that we saw in that period.

Today, although I am more confident that the past practices have been stamped out, a need remains to ensure any new business practice continues to operate on the ethical side of the ledger. As we have seen, the office printing and document imaging industry has had to continually reinvent itself and fight off the hangovers of the past. Many times there has been suspicion from customers who have been on the wrong side of unfavourable contracts. Other customers have felt that there was a large gap in what they thought they were receiving and what they actually received.

Without trying to defend the industry, my view is that their problems were most likely caused by a few people inside a few organisations where the business culture conspired to promote poor behaviour.

THE FIRST SIGNS OF CUSTOMER DISTRUST

My earliest recollection of signals that the industry was not operating in the most professional or ethical manner was in the very early 1990s. Rumblings came from New Zealand where the photocopier market was dominated by a New Zealand icon brand, a non-traditional manufacturer that distributed a well-known copier brand at the time.

The company's sales teams were highly successful at selling their product solution to the customer. The solution was a financed program known as a blue chip plan that could incorporate elements such as the photocopying hardware, the service and maintenance, the toner consumables and paper supplies. It appeared simply as a page price rather than a large capital asset cost. It appeared too good to be true. In many cases it was. There was a sting in the tail and that sting was not always visible to the customer. Serious Fraud Office of New Zealand investigated this company but no charges were laid. But what this did do was put the industry under the spotlight both in New Zealand and in Australia for several years. Many salespeople would tell fictional stories about their opposition's tactics, which only reinforced the air of unprofessionalism in the eyes of the customer and fed distrust in customer and industry players.

THE SAME BUT DIFFERENT

Although the financed product solution of the 1990s has gone through a number of iterations it has matured into a product of choice to many sales organisations around the world. The financed product solution today is known as a page per print contract. Prior to digital, when analogue devices prevailed, it was more commonly called the cost per copy contract. Many sales companies had variations with their own names, such as the blue chip plan, and their own variations on how the product functioned. They all looked similar but not all products worked the same way.

The New Zealand company above had its own version and it appeared very creative to the customer at the beginning of the sales process, as well as very financially rewarding for the sales-person and company. However, it came unstuck as the customers came to the end of their first contract term. The weight of the ballooned costs at the end of the contract was unexpected and too large for some customers to absorb. In many cases the debt was greater than the remaining asset value.

Customers only became aware of this predicament if they were considering a change of suppliers or companies. If the contract simply rolled over, they were none the wiser. In some cases the payment costs may even have gone down or stayed the same on renewal, making the customers think the deal was a good one.

BE AWARE OF THE BUNDLE

On the surface there is nothing wrong with bundling products and services into a pricing model. It works financially for both the customer and the sales organisation – within reason and when it's appropriate. Some finance products certainly depend on the customer's situation at the time and it certainly requires full disclosure on behalf of the sales and finance company providing the product. However there is a risk when this practice is exploited.

Best Practices

In my view, it is best practice to ensure my customers always understand how the contract is broken down and how it has been put together. I have assisted a number of large Australian customers to un-wrap their existing contracts and discover that they have been over charged (or are overpaying) on a regular basis.

In some cases the customer is underpaying and on the surface it appears to benefit them, but this later proves to be wrong when the supplier requests the shortfall to be paid. This can be embarrassing for both parties if the financial year's books have been signed off. To ensure that the business relationship is maintained I always strongly recommend that this customer reconciles any outstanding costs with the supplier as soon as possible.

In another case the customer was still paying their all-inclusive click or print charge despite the minimum print volumes for the month having been achieved. At this point the excess print charges (prints produced after the minimum print volumes have been achieved) should have reduced to the service and maintenance print rate only.

CHAPTER 7:
A FINANCING REVOLUTION

PART 1: THE PAGE PER PRINT SECRET RECIPE

Here is an example of how a page per print (PPP) contract is put together. There is some more detail about this in Chapter 21 but right now this is all you need to know. This example shows how excess prints can work against a customer. The prices are not indicative of the real world because I have rounded them up and down to illustrate the point.

Ingredients
- The capital price of the hardware (total cost of capital to be financed), say: $10,000
- The rate for servicing which, in this example, is a flat rate that includes all onsite service and maintenance including spare parts and all toner consumables represented as a pure click rate charge, say: $0.008 cents per black and white page and $0.080 for each colour page

- An agreed minimum monthly print/copy volume to be achieved (across the total fleet or on each device), say, for this device: 5000 prints per month
- The contract term, in this case: 48 months

Method
1. Amortise the capital over 48 months
2. Run all the costs through an equation to determine the price per print
3. Present the deal to the customer

In this case, to make the math simple, let's say we worked out a flat rate of $0.20 per page as the all-inclusive page per print price. And, to keep things simple, let's say colour and black and white pages are the same price (despite colour usually being more expensive in the real world).

The customer will pay $0.20 per page up to their minimum monthly print volume of 5000 prints. The customer does not have to achieve 5000 prints per month. The minimum refers to the minimum copies they will be charged for, regardless of how many they print.

However, once the print volume goes over the minimum 5000 prints mark, the price of the additional prints could drop because the price per page includes the amortised capital cost of the machine, or the term could reduce because the contract is for a set number of pages. Over 5000 prints the customer should only be paying the maintenance cost of $0.008 per page for black and white and $0.08 for colour, or the term could be reducing from 48 months to a shorter period.

Depending on the contract, this price or term reduction does not happen automatically; as was the case with one of my clients. We held a contract review and established that they were overpaying and were entitled to pay only the service charge: the

$0.008 per page for black and white and $0.08 for colour.

My customer was able to recoup a substantial refund and use their position to negotiate a better deal because it was based on a substantial amount of devices and print volumes of over one million prints per month. Multiply this over a 48 month contract and the savings are significant for the customer.

Improvements in pricing and contract management have improved the accuracy of PPP contract management so they are more transparent than before. PPP contracts now account for approximately a quarter of corporate business contracts in Australia and New Zealand.

Other variations of these types of finance contracts include:

- Term-based contracts with a fixed price for the term and that expire when the term is complete
- Volume-based contracts, where the total contacted volume is fixed and the term may reduce if this volume is achieved earlier than expected

Why the customer doesn't trust the industry

One of the practices that afflicted the industry in the early 1990s was recommended retail pricing. Recommended retail was more "flexible" or loose than it is today. In past years, prior to the establishment of the Australian Competition and Consumer Commission (ACCC) in 1995, recommended retail or retail spread was usually an inflated number – probably where the term "blue sky" came into effect.

Prior to the ACCC, the industry was able to inflate or manip-ulate the recommended retail price paid by the customer when they chose a finance arrangement to acquire the asset. This caused customer concern about over capitalisation and high debt exposure at the end of the financed term.

How this worked was that a salesperson could fund an existing payout of a device as well as a new device in a new financed arrangement. The only provision is that both the payout and the new asset had to be within the total new list price of the new goods. As the list price was usually very large, financing became very attractive and easy to accommodate.

Therefore, as part of the sales process, a salesperson would ideally position a device that was more designed to suit a client's needs and requirements. However, on some occasions, this was driven by the salesperson's knowledge of which device had more retail spread. If this sale was successful it provided the salesperson with a healthy commission cheque and also meant that the client's new monthly rental included a lot of "blue sky" or "air". This practice has been abolished in Australia by tough retail pricing regulations and compliance.

Hidden traps still exist

Today I still see practices that concern me, such as the automatic contract rollover. Customers talk to me about two types of contract rollovers: the finance contract rollover and the service contract rollover. Both infuriate customers when they discover how this process works, especially when their contract has been rolled over and the sales and finance organisations stand behind their rollover policy as if to absolve them of past wrongs.

A finance contract rollover can automatically renew after its term for another month, three months, 12 months or more, and continue renewing, depending on the contract terms and conditions. Over a decade or so ago the finance contract could roll over for the same period as the original contract term. For example, a 36 month contract could renew for another 36 months. A service contract would rollover the same way as the finance contract.

Automatic rollovers can usually be stopped by providing formal written notice. In many cases, verbal communication is not enough. Some contracts require the end customer to make the written notification 28 days ahead of the end of term, others 90 days prior. If notice is not given the contract simply rolls over and a new contract period commences.

In the past these rollover clauses existed to lock the customer into the sales or finance company. The rollover would have no impact if you wanted to upgrade with the same sales or finance company. However, if you tried to cancel the contract (mid-term or prior to end of term) you would attract penalties that may have made it too expensive to transfer from one supplier to another. In these cases the customer would just roll over and be prepared next time.

PART 2: PAGE PER PRINT CONTRACT WARNING

Today my largest concern lies in how new devices are added to existing page per print (PPP) contracts and the way those PPP contracts may tie the client into future problems.

Sometimes these deals happen at the request of the client and sometimes they are driven by the supplier. While they may look good at the time, I spend a lot of time trying to manage the impact of these deals on my clients long after the deal was done. Unfortunately, by the time the customer realises they have a problem, World War III is often about to break out between supplier and customer.

Here is one example of how it can work:

1. The salesperson receives a call from their current client wishing to add three new devices to their contract because they have a new business area and need more equipment.

2. The salesperson visits and explains that they are currently on a page per print (PPP) program with the flexibility to add devices at any time. Sounds good so far, right?

3. When the client asks how it works the salesperson says something – in a very simplified scenario – like this, "You are paying $0.16 per print for all black and white prints and adding the three new devices will only increase it to $0.17."

Let's explore what is not being said:
1. Somebody has to pay for the hardware – in this case it is most likely you, the client – and the capital will need to be financed over a term, e.g. 48 months.

2. The term depends on how the contract is structured. The contract for the new devices can be done in three ways:
   - Create a new contract for the new devices, with its own term and pricing structure
   - Create a new contract for the new devices that is adjusted to finish at the same time as the existing contract
   - Add the devices to the existing contract and extend the term of the existing contract so that your existing devices pay for the new ones.

3. This is a PPP contract so there is a minimum print volume and the monthly print volume can be attributed to each device or to all devices in the contract. When the salesperson adds the new devices to the contract they can increase the print price across the entire

contract and may also try to increase the minimum print volumes. Again, your existing devices may be paying for the new ones.

All three variable components can be used on their own or in conjunction with each other, depending on the transaction, the salesperson, the sales organisation and the financier. The main point is that minor increases in the page per print charge, term extension and/or the minimum monthly print volume can generate substantial additional income for the sales vendor and put you at considerable risk.

The risk lies with the customer when new devices are continually added to the all-inclusive page per print contract. I have seen, in far too many cases, a third of a fleet made up of aged equipment, minimum monthly print volumes continuing to increase, service rates continuing to increase to pay for each and every new device going in. The contract becomes larger and larger. The term extends and extends and the print volumes increase.

Every salesperson will see you as a prime target, but the incumbent is usually a protected species as they have better visibility of your real costs (not just the stated click charges) and, in most cases, a lower financial cost (payout) than their competitors. I have seen many irate customers realise that the transition cost to change to an alternative supplier has brought them full circle to the devil they know: the incumbent.

By the way, the incumbent bets on their home-court advantage and usually wins this game. So in some respect a page per print contract is a win-win for the sales company and the salesperson. In fact it may be better written in some cases as lose-win – the customer loses while the supplier wins.

My final note on this is that not all supplier organisations have this view. I have had the good fortune to work with some very sincere, hardworking and professional organisations that also absolutely detest this malpractice of salesmanship.

A PPP contract can be very beneficial for the customer if it is negotiated well, with an eye for the small print. However, negotiated badly, with the fine print ignored, the PPP contract can cost you money and opportunity.

Dear Customer: No, we don't recommend it

If there was one aspect of selling that I could change it would be around the sales team understanding when an all-inclusive PPP contract is not the most suitable for their client.

Not all customers are well suited to an all-inclusive page per print contract. The salesperson, the sales organisation and the customer need to know when a PPP contract is not right. I do not believe that, in many cases, the sales teams really understand, or have been trained to understand, when and when not to use this type of all-inclusive contract.

The process of qualifying a client at the business level must clearly identify the business the client is in and how they plan to expand or condense their business operations. The salesperson must understand the core essence of their customer's business and their plans for the next three, four or five years. Is the client on a growth trajectory? At what speed will it grow? Is the client thinking of downsizing? Are they an acquisition play business or looking to acquire new business shortly?

Sure it may be difficult to crystal ball what the client may do, but at this point there is a valid reason and a professional responsibility on the salesperson, sales organisation and financier to at least pose these questions as it is my belief that there is a professional obligation to set out, for the client, the risks and benefits to them of term-based and volume-based contracts.

Establishing a minimum print volume locks the client into a commitment guarantee. As I have said before, print volumes that exceed the monthly minimum may help to reduce the real term but the client still needs to fully understand their contract obligations if they plan to make significant business changes such as divesting a major business division.

Let's recap the all-inclusive page per print contract:

1. Set and agree a minimum print volume with the customer. This can be based on the term of the agreement and is usually a monthly print volume commitment calculated by multiplying the monthly print volume of the device fleet by the term of the contract agreement. For example:

   500,000 prints per month X 48 month (term)
   = 24,000,000 prints over the term

2. Choose between a volume-based or term-based contract (there are other hybrid contracts but for simplicity's sake we'll stick to just two). Remember both contracts use minimum print volumes and are calculated over an agreed term. The major difference is that the volume based contract can run shorter than the agreed financed term as any excess print volumes reduce the time taken to achieve the minimum agreed volume.

   Usually when a customer signs a volume based contract, prints (colour or black and white) that exceed the minimum prints are still charged at the full page per print rate and do not reduce to the service and maintenance only click rates. If the contract is not varied, the client is committed to this volume as a minimum. *Note: some finance products operate differently and you'll need to gain independent industry advice to find which product is best for your business.*

3. Your minimum monthly cost is calculated by your all-inclusive print (click) charge rate X the minimum monthly print volume. For example:

500,000 prints per month x $0.16 cents per print = $80,000.00 (assuming black and white prints only)

This price includes the capital and service and maintenance costs to support the fleet at this minimum monthly print volume and excludes any prints over and above the agreed monthly volume of 500,000 prints. Also, if the client prints less than the 500,000 they are paying for they still have to pay $80,000, although there is sometimes a quarterly or annual reconciliation process to balance the books.

Now let's look at how being locked into a PPP contract can work against a customer.

PART 3: BEST PRACTICE RECOMMENDATION FOR PPP CONTRACTS

The problem

The customer currently owns and operates three different businesses with a total fleet size of 200 devices. The business is currently under a PPP program with a minimum volume commitment each month across the fleet. The client has two years remaining to run on a four-year contract. The client has decided to sell one of the three businesses to another organisation.

The business that is being sold is currently responsible for 50 of the devices and 200,000 prints of the minimum 500,000 prints per month.

As the owner of the all-inclusive master agreement, the client (the owner of the three businesses) carries the obligation for the total print volume, regardless of where they are produced or who produces them.

If you are in this situation, you will need to manage one of two scenarios:

1. Ensure that you can transfer the print volumes and assets to the new business owner. This will allow you to reduce your total monthly print volume commitment in accordance with your new actual usage once the business unit is sold. This needs to be included in the commercial arrangements when selling the business.

2. If you do not transfer your print volumes including the assets you, as the contract holder, will still be paying the total minimum monthly print volume even if the assets are now in a business that belongs to someone else. This can occur because page per print contracts can be difficult to break out of and print contracts are usually a late consideration in a commercial asset sale.

My recommendation in these instances is to consider an audit prior to finalising the business sale so the appropriate devices (and related print volumes) are transferred more effectively or at least taken up as an ongoing cost consideration.

I know my clients do not want to pay for print volumes that are not being produced each month because the buyer of the new business unit already has a national agreement with another vendor, or chooses not to take over the contract, device or volume commitment.

I have been brought in by clients to ensure that their existing business does not get lumped with assets that are not right for the business, such as those that are not cost effective, not the right functionality or near end of life, as part of an acquisition.

I believe that sales organisations must be more open about the downside, or risks, of an all-inclusive page per print contract based on minimum monthly print volumes. I see far too many sales presentations, by novice or inexperienced salespeople, stating the ease of calculating a simple print charge rate. They overlook, or are unaware of, the negative aspects of such a financial contract.

In a market place that is becoming more fluid in its structures and business models, and with global and local markets fluctuating so quickly, the all-inclusive pay for print charge finance product needs to be reconsidered in the light of rapid market change. Organisational resizing now requires more planning in both personal and operational agility. Contracts that lock-in the business over a long term may not be the most desirable product in the future.

Best practice recommendation

I recommend that the salesperson, in conjunction with the sales company, have the end client sign off a disclosure document that says the client has been informed of the risks and obligations of any all-inclusive page per print contract or contract variation. This provides evidence that the sales organisation communicated both the advantages and potential risks to the customer of the all-inclusive page per print contract. This provides an increased level of confidence to the next executive that manages the contract that the vendor or sales organisation acted in a transparent and professional manner.

From a customer perspective, in addition to what has been explained I also recommend that there is a print volume reconciliation process that helps to manage print volumes across high and low print volume months and devices.

Section 3:

THE TRANSITION FROM PRODUCT-LED TO SERVICES-LED

CHAPTER 8:
THE FIGHT AGAINST COMMODITISATION

WHAT IS COMMODITISATION?

Commoditisation is the process by which goods with economic value and distinguishable attributes (uniqueness or brand) become simple, interchangeable, commodities in the eyes of the market or consumers. It is the movement of a market for that product from differentiated to undifferentiated price competition.

From the business directory (www.businessdictionary.com) comes this definition; "Almost total lack of meaningful differentiation in the manufactured goods. Commoditised products have thin margins and are sold on the basis of price rather than brand. They are characterised by standardised, ever cheaper and common technology that invites new suppliers into the market to lower the prices even further."

Manufacturers or service providers that sell and market office printers and multifunctional devices are very sensitive about commoditisation. Most in the industry think they provide a unique product and I support this view to a degree. However, before we go further it is important to clarify a few points:

First, parts of our industry – the office printing and document imaging industry – are certainly and clearly commoditised. Toner sales and the hardware-led transactional sales are particularly commoditised. Customers are inundated with sales companies and end up receiving six or seven proposals from different organisations trying to sell them a multifunctional device or printer.

Each quote includes a service price that is clearly at or below commercial floor. I call this "quote and hope". I have seen this happen to businesses from two-person organisations up to large multinationals. How can the customer know who they should be dealing with when they have salespeople representing the brand directly, sales agents who represent a brand indirectly, resellers, IT integrators and others all selling the same brand of device at different prices?

The highly competitive nature of the industry drives and accelerates pricing erosion. This looks great for the customer and not so good for an industry that resists the idea that it is a commodity business.

Customers will and do see this as a commodity product. When it is hardware-led, price becomes more important in the conversation, and that leads to a price war between brands and suppliers as they try to maintain or build market share. Market share, in turn, provides unit (device) numbers back to the manufacturer's parent operation and feeds the economies of scale for manufacturing plants based in a number of geographic regions. As you can see, the industry is caught in a macro and micro-structural and cyclical process.

The difficulty with such a process is that it's hard to break the cycle. Imagine swinging through the jungle, swinging from one branch or vine to another. It's quite easy when you see the other branch and vine ahead, so letting go is not always the problem – you can see what is about to happen. Now imagine if you had to let go before you could see the next branch. Would you let go? That comes with higher risks.

That's what the majority of the office printing industry manufacturers are facing right now. It probably comes as no surprise that the majority of product brands (i.e. consumer electronics) that produce (manufacture) office printers and multifunctional devices are Japanese. The exception to this is brands such as Samsung (Korean), Pantum (China), Hewlett-Packard, Lexmark and Xerox (all from the USA).

Note: Fuji Xerox Australia is an affiliate of Fuji Xerox Co Ltd, a joint venture partnership between Japanese Fuji Film Co (75%) and Xerox (25%).

These brands, in particular, are facing more rigid resistance than ever before. Their challenges are not temporary. We are seeing global structural and economic changes that are forcing many economies and industries around the world to rethink their operating models. With the increasingly rapid change in technology, old business practices are quickly evolving and this has a direct effect on industries like office printing and document imaging.

With this in mind the industry is attempting to evolve without killing off their existing business. It's a delicate balance that they have to get right. If you're too quick or too slow to change you may not be in the strongest position when a new market develops or when a stronger market leader can take the right advantage.

To manage this balancing act, office printing manufacturers across the globe are trying to protect their manufacturing volume while preparing for the new world. Transactional-led sales support the volume manufacturing model of the industry. The new services-led engagement supports their latest movement into MPS.

A true MPS offering should be less about device and brand, and more about providing a business outcome. This means that device volume units will potentially decline and with the advent of print page volumes declining, this means that toner consumables (the very heart throb of the business) may start to decline.

The second point I'd like to make is that managed print services should be a services-led business model. As such, I would have expected it would not fall into the commoditised play as quickly as it has today. However this does not mean that managed print services are not being positioned correctly or delivered as planned.

Nonetheless there are some aspects of managed print services that I continue to come across and witness first hand which are represented more as a transactional led sales process. I think it is because salespeople are being incorrectly led by their organisation and at the same time have fallen back into their old bad habits, such as:

- Salespeople becoming desperate for the sale as they are still are measured using device or box units
- The sales process is not consultative because the salesperson has a clear end game before they start
- The sales force has a low skill level because of a lack of training
- The salesperson is not having the right conversation with the right people
- The customer is ahead of the game or understands the game better than the salesperson
- The client drives the conversation to a commoditised viewpoint as the customer's perception is that the product is low on their priority list compared to other business areas or IT functions
- The client thinks the brands or devices of the services provision are poorly differentiated between the competitors
- Any combination of the above.

CHAPTER 9:
WHEN THE CLICK RATE PRICE GIVES YOU A SEAT AT THE TABLE

In many cases in Australia a solution is sold on a price point. The reason I say this is twofold.

First, I have been in many meetings as an IT advisor and witnessed the end client's commercial negotiation and the sales engagement from the supplier. It usually goes like this:

1. The end customer tells the supplier that their service and maintenance print (click) charge needs to be "xyz" to have a seat at the table. "If you can't meet this then we won't be doing business with each other."

2. Now all credit goes to the opposing side, the sales-person speaks extensively and with great authority, explaining how quality, value, brand, service, reputation, responsiveness, optimising workflow, business process improvement, contract management and managed print service experience are all worth

more. The salesperson has convincingly restated their position.

3. The end client, without flinching, says, "I appreciate you have all what you say you have, however I am receiving this pricing from a number of your competitors. More importantly, I have extensively researched the market and have spoken to my many colleagues. I have discussed my plans with my strong peer networks that are in the same role as me. I have read many blogs on this subject from many players and experts across the office printing industry."

4. The client strikes the final blow, "After doing all this research, I really cannot see that the risk of going with you or another leading brand is that high. I do not see much difference between you and some other brands." A quality salesperson doesn't bad-mouth the competition (it makes the whole room cringe) so they stay quiet here. Then the client says, "I know what you are offering, in your mind, may add value to our business; however you need to understand that we need to partner with a business that can come to the table in the first instance."

Now we are in the perfect standoff. The end client says nothing. The ball is firmly at the feet of the salesperson, or their managers if they are with them on the call. What to do? They can continue to negotiate up and continue to establish value (which is OK up to a point) but the client has made it clear where they stand. Who is in control?

In the end the salesperson or their manager has limited options. In about 90 per cent of cases, we see the manager – surprise,

surprise – agree to the service rate that the customer asked for at the beginning of the meeting. The remaining 10 per cent walk away from the deal.

Either way, the client says thank you, shakes hands and the salesperson departs.

The client has not lost anything as in most cases they get what they want and usually have most of the brands or sales channels going through the same process so it becomes relatively common. They generally establish the pricing boundaries and the pricing floor the market will drop to. Then it's just a process to establish who is in or who is out.

I appreciate that this illustration appears very simplistic and in many cases it is. I feel for the salespeople (I know firsthand as I have been in this same position myself in the past) who have to deal with this process as some brands, products and value added services can be clearly differentiated from their competition. However the client has gained a significant baseline start and it's one that they did not have to work hard to achieve.

It also depends on who you are selling to within the business (for example IT, finance) and how you have developed and built your relationship. If you can establish a level of trust with the client through your consistency, your value proposition may not be driven down as quickly or as low as it could be. But if you are coming into the process late, your ability to negotiate up is very difficult indeed.

We often see suppliers and salespeople who come into the process late and lead with price because they have not been able to develop a relationship with the client. Usually when price is the lead, the ploy is to low-ball it. This is done to either make the customer second guess what the other suppliers have not given them, or it's to hurt the opposition. In a sales process, if you are going to lose the deal, you may as well make it difficult for the opposition along the way. If the opposition wins the account, but has to match your price, it will cause them financial pain. Yes, it can be a vicious business.

The second reason I think the industry has developed sales pricing as their vehicle of choice is based on the sales skills within the industry today. Many experienced sales teams face the challenge of competition that quickly pitches price levels to the client, such as the service print charge rate they offer.

These situations are embarrassing and comical as the novice salesperson barely introduces themselves before telling the client what their service print charge rate is.

Many experienced salespeople who used to be in the Australian industry one or two decades ago, have left though natural maturity or transition to a new industry. Unfortunately the industry as a whole did not continue to train and develop their salesperson as they moved through. Mostly the industry can't afford to train new people coming into the business. The cost of sales is already too high for many players today.

By the way, direct and indirect manufacturers and many resellers do not want their salespeople to have an in-depth understanding of the office printing and document imaging industry. Many of the better performing sales companies are employing more professionally skilled and commercially orientated talent. They look for people with skills in a discipline such as IT, finance or some core proficiency in a vertical industry such as manufacturing, insurance or banking.

Today we see many sales organisations clearly breaking their sales teams into roles and groups across varying market and industry segments. For example many sales companies employ account managers who are responsible for managing and coordinating the varying activities across their customer's environment and business development managers who are responsible for identifying and engaging new business with the aim of securing the client's business. Specialist roles are also used and are varied within the sales organisation.

They are used as knowledge experts in a particular field, such as software specialists or colour specialists. In addition to their unique skill sets and experience they allow account managers and business development managers to maintain their direct focus on the customer's core requirements, increase the efficiency across the account and reduce distraction.

Specialists are also used to identify potential new business opportunities that the sales teams and organisation may not identify as easily. When a sales company launches a new product, market or services opportunity they bring in a specialist as a way to incubate the knowledge and learning, and transfer the skill to the rest of the company. This process helps to speed up the sales activity and results while reducing risk and distraction.

This has led to an industry with a wide range of sales capability in its ranks. As an industry, however, it continues to have a relatively short-term view on securing clients and ongoing client retention. This may be partly due to their existing approach of reward and recognition for the sales teams. When you take this one step further you realise that the current measurements for success are still very much aligned to box (device) volume unit activity.

This focus on their remuneration and incentives has ultimately produced a certain sales behaviour that appears more mercenary in approach. I expect that the industry still believes it needs both types of salesperson – strategic solution salesperson and box movers – especially as it tries to transition from high volume "box" transactional sales to a new services-led business model.

This latter group of salespeople continues to approach its business and sales opportunities purely as a numbers game. They cannot necessarily invest the time to develop and cultivate relationships as they are incentivised on "box" sales activity. Not necessarily their fault. That is how the industry rewards their activity. In part that is what the local subsidiary requires to maintain the pricing rebates they receive from their parent organisation. If these device numbers or box units are not achieved according to plan then pricing for devices may increase, impacting customers and potentially eroding future device unit sales numbers.

In other cases, the local executive will be under pressure to hold their forecasted positions to make their local revenue and unit targets.

On a recent visit, Stuart Jacobs, Regional IT manager of Sims Metal Management, with more than 300 devices nationally, highlighted this point well.

"When you consider the supplier's reputation as being a very important requirement to winning or obtaining the business, what you don't want to have is a business development person who continues to move "ship" from one brand or sales company to another every six months. Then he has the audacity to try to sell you something when he doesn't know your business and what you're trying to achieve. This looks bad for the business and most importantly their reputation.

"Once the business is secure, we want an account manager so we don't have to deal with the problems directly."[4]

This inconsistency is very damaging and concerning because as a customer you do not want to have someone managing your account and then disappearing to the next job.

In Australia we have a smaller quantity of professional business agents in and across a number of states who conduct their business in a very proficient and prudent manner. They have built a reputation over a long period and have operated their business like any commercial entity. They have built relationships through longevity and trust. These are usually either a direct sales agent for one recognised brand or they are an indirect channel model supporting multiple brands and offerings.

Outside this we continue to see new players enter the office printing and document imaging market with fewer skills and overall less commercial office printing and financing capability. Some are entering from an existing low-end commoditised market position such as toner supplies or office products suppliers. Although there are certainly synergies and alignment that make great commercial sense for the customer, these newer sales organisations may be increasing the commoditisation cycle in Australia today.

CHAPTER 10:
AN INDUSTRY UNDER PRESSURE

AN INDUSTRY AT ITS CAPACITY

Which office print manufacturer will be taken over next and by whom? It's a big question and it's always a hot topic. It's one that we have been asking ourselves more frequently over the last few years. More recently we have seen Canon Inc. finally acquire Oco (the Dutch multinational) in 2012. While Kodak entered chapter 11 (bankruptcy) in January 2012 and re-emerged some 20 months later in September 2013. Although both brands were predominately from the production publishing side of the coin (not your typical small office device), both are examples of industry consolidation and in Kodak's case failed to see and make the necessary changes to adapt to a new fast moving world.

The impact to the industry when these changes occur is usually felt at both a local and global level. In fact we continue to see many acquisitions taking place all around the world including some in Australia. In truth there have been so many over the last few years it's too hard to single out specific purchases in this book. However at this stage what we are seeing at present is more acquisitions rather than mergers and this is reflected around the world.

When we examine these acquisitions we see some are within the scope of the office printing and document imaging industry while many are from parallel industries like IT. In some cases they are from further removed industries. This may indicate a change in strategy, business model direction and customer markets for both existing and new entrants.

However, if we take a look within the office printing and document industry specifically, we can see that over the last 25 years or more, the number of office printing manufacturing brands has reduced by nearly half. Today around 16 brands manufacture, market and sell some form of office printing (copying) output devices across the world and occasionally new brands, such as Memjet and Pantum, come to market. These 16 or so brands manufacture both A4 and A3 MFD or MFP devices across laser, ink jet (including new page wide array ink jet technology), and bubble jet for the business and home office environment. Brands that are owned by other brands have not been counted.

I do not include commercial printing, 3D printing, labels printers and screen-printing as they are entirely different markets. 3D printing warrants a book of its own.

NEW REGIONS FOR GROWTH

More recently there has been increasing pressure on the office printing and document imaging industry to grow market share and revenue while taking costs out of the business. In Australia, cost cutting is particularly important; while India, China, the greater Asian region and South America are seen as large market opportunities for the office printing industry. The hope is that the plateau, or even decline, in developed markets may be offset by growth in the developing markets, although timing that perfectly will be difficult.

Another factor in emerging regions is that they may not want to adopt past and existing business models and the business practices of the West. Emerging market economies may consider it a greater risk, costly and somewhat uncompetitive, to adopt what they believe is the old-fashioned technology platforms of their western counterparts.

Many medium to large enterprises that currently operate in western societies have been building their business models, processes and technology infrastructure to meet a marketplace that continues to rapidly change. More of a concern is that many businesses here in Australia and in other western economies are struggling to dynamically change and adapt to new customer-driven opportunities and markets. The technology that underpins their current business is no longer agile and suitable for today's fast-moving business and information-intensive consumer society. Therefore the previously invested technology becomes somewhat of a legacy and an inhibitor of change.

New regions such as China, and their success, may depend on their accelerated technology adoption and innovation. In other words they have a "green field advantage" of technology deployment – no legacy to live with. Their aim will be to not adopt existing processes and technology that their western counterparts are struggling with.

I expect that they are not just aiming to be the same or somewhat competitive in the global market. These growing economies clearly understand that they have a unique window of opportunity.

With this in mind businesses from these growing economies and regions may not take up the device unit numbers the office device manufacturers are hoping for. This may limit the speed of growth for many of the office equipment manufacturers world-wide or reduce the supply opportunity the manufacturers require to sustain their businesses or manufacturing in the future.

THE PERFECT STORM

More recently we have seen even stronger signals that further industry consolidation may not be far off for the office printing and document imaging industry. This industry consolidation is moving closer due to a number of competing forces, including:

- Economic structural change: a manufacturing-based economy transitioning to an information-based economy
- Increasing acceleration of globalisation
- Shifting demand and supply.

Since the global financial crisis in 2009, we have seen global markets come under increased pressure to fund new and ongoing financial commitments across both the public and private sectors. This global challenge has reduced business and consumer confidence and continues to have a direct impact on the office printing and document imaging industry.

Additionally, natural disasters such as the earthquake and tsunami in Japan heavily impacted a number of industry brands. The delays in devices, component parts and deliveries of supplies around the world proved to be costly across the manufacturers' supply chains.

At the same time we started to see office print volumes slow down and for the first time in history we have seen office printing volumes go into decline.[5]

We expect much of the office print decline occurred when

paper-based back office functions migrated to digital workflow, largely eliminating the need to print. These workflow applications included such items as human resources request forms (such as annual leave, sick leave, salary receipts, and internal requisition forms), business processes (such as mortgage applications and associated documentation for home loan approvals), accounts payable, accounts receivables and more.

In addition some would argue that, due to the economic conditions resulting from the global financial crisis, MPS has become more effective at reducing and consolidating the client's office print fleet. The reason for this view is that the global financial crisis has promoted a cost savings discipline, focus and acceleration for most businesses around the world. Many MPS firms have continued to promote cost savings of up to 30 per cent which has ultimately attracted the customer's attention.

Although device rationalisation may be having some impact in reducing print volumes within an office environment this, in itself, would not be the major reason for its decline.

IF YOU CAN MEASURE IT THEN YOU CAN MANAGE IT

One of the breakthroughs that supported the growth of MPS in that early period was the increasing availability of independent monitoring software tools. These tools could interrogate networked printing devices using simple network management protocol (SNMP), an internet-standard protocol for managing devices on IP networks.

Due to these SNMP agents generating very little network traffic, being quick and easy to install and not handling information that required a great deal of security, they were easily accepted and installed on many clients' networks.

The typical information passed across the network by these SNMP agents is known as the machine information base (MIB) and includes the device serial number, quantity of colour and black and white prints it has produced, the toner usage, whether the device is on or off the network and the status of various features and functions (such as if the paper tray is empty). No real information about the client's business, what or whose document has been printed is visible or recorded.

This technology provided two compelling opportunities. It allowed:

1. different devices from different brands to be managed by one independent tool

2. all independent providers to compete directly against an OEM brand because the customer was not locked into an OEM's preferred monitoring tool.

Before these SNMP agents became available the smaller independent dealers or service providers were limited in their capacity to support a customer's fleet unless they were a dealer or partner of the brand in question. In these situations the OEM brand had the power in the relationship and owned the customer and their information.

Conversely, some OEMs were also somewhat restricted by only being able to read their own device's MIB and not others, negatively impacting their capability to proactively manage a full independent services offering.

The new monitoring tools evened out the ledger between OEMs and independents.

In addition to the SNMP device monitoring tool, we were introduced to client-based architecture that could track local (non-networked) devices to a limited degree. The real value of this client-based tool was that it could track every networked print job via the user's workstation. This was ideal to track user printing behaviour, control and limiting print usage (for example by allowing black and white printing only), redirecting print jobs to a more cost effective device, tracking print traffic, limiting big print jobs (such as internet printing) to quiet times and printing from applications like Microsoft Word, Excel, and PowerPoint. By monitoring the device and print jobs you knew what time the prints were being produced, by whom and how much they printed in a hour, day, week or month. It is important to note you could see the file name of the print job, but not the content or pages of the document.

Being able to see and report on document file volume usage turned up some curious statistics. For example, the most common colour document printed in 2010 across corporate Australia was *Jamie's 30 Minute Meals*. I imagine they were generally printed overnight – after most people had left for the evening – and I wonder how many of the finance departments or divisional managers signing off print costs questioned the spike in internal printing costs at the time, especially their colour costs.

On a more serious note, I was asked to audit a large customer who could not account for a sudden, unbudgeted increase in colour printing costs. I undertook research, analysis and interviews and saved my project sponsor some embarrassment. The CEO had mandated a three per cent increase in top line revenues while maintaining current cost levels. The sales and marketing team went into overdrive, producing more tender response documents, which relied on colour as a selling advantage for their product offering. The tender responses were printed in-house and caused the dramatic increase in colour printing costs.

I am pleased to say that the CEO's mandate was successful and the sales team secured their largest ever contract with a major utility that had over 1500 office locations across Australia. The three per cent revenue increase was achieved, but in doing so pushed up operating costs. However the insight (or hindsight depending how you see things) is that when the CEO requested the revenue increase there was no way, at the time, to project the additional cost of such an effort. The ability we have now to project these costs could have prompted the CEO to ask for a four per cent increase to covering the increased operational costs.

Providing more effective software and data tools, to mine and map actual real print volumes and costs at user and device level, gives us the ability to provide greater transparency to the end clients. This has resulted in clients realising their real costs of print and the impact that colour printing, for example, has in terms of their costs.

My research and extensive client audits using data tools and data mapping based on typical customers with more than 100 devices, have yielded the following insights:

- Customers print black and white to colour on a 75:25 split; that is 75 per cent black and white print versus 25 per cent colour.
- Customers costs are the reverse: they spend 75 per cent of their budget on colour and 25 per cent on black and white.

Obviously these numbers vary across industries; a retailer would only produce black and white prints for internal use while a sales and marketing business such as a commercial property agency or an advertising company may use colour extensively for proposals, flyers and brochures.

When clients see this level of visibility they quickly realise where their costs are really coming from. The saying, "if you can measure it, you can manage it", really rings true in these meetings. The client then is more than willing to put in more accountability to manage their print volumes.

When a client actively manages print, print volumes tend to drop dramatically, before plateauing at a lower level. This would appear to be a natural effect when moving from an unmanaged environment to one that is managed and aware of its printing footprint. For example, in my experience, when we communicate to office users that we are now monitoring who prints what and when, print volumes drop by at least 20 per cent.

Furthermore, my experience shows that providing a "before and after" change, both physically and financially, in respect to energy and carbon usage or paper and consumables supplies, most customers are very satisfied. Even more so when I illustrate further environmental considerations by breaking down the key elements and then rolling this back up to provide the business with metrics so they can either tick the box on set objectives or obtain accreditation for compliance, corporate governance and triple bottom line reporting. Once you start to bring in this wider view the clients realise managed print service makes much more strategic sense than they first realised.

OVERALL PRINTING IS INCREASING?

On this point I would like to make it very clear when we consider office printing as a term, we have been very specific to ensure it refers to printing within an office environment. Printing in a broader sense includes commercial or industrial printing and packaging. In real terms, I dare say, printing has increased overall; mainly through increases in the area of print labelling and print packaging.

The internet has proved to be a growth engine for internet retailing. This has meant businesses now pack and ship more products globally than at any time in the past. As a result, print packaging and printing labels have thrived. In October 2013 iTWire said, "Australia Post has announced record financial results. It

ascribes the numbers to a continued growth in the digital economy – delivering products ordered on the Net. Australia Post's
parcels business generated $355 million profit in the last financial
year, with domestically delivered parcels volume growing by 9.3
per cent, fuelling underlying revenue growth of 8.4 per cent."[6]

If one was to suggest that overall printing is increasing, based
on the iTWire example, I would have to agree. However the
opposite is true for printing in a general office environment. This
is certainly being ratcheted down and it will continue to decrease
over time.

In addition to declining print volumes, the office printing
industry has been impacted by the global financial crisis, an ever
increasing economic structural change in virtually all economies,
and a number of natural disasters that have impacted the industry's supply chain. Now we have the increasing momentum of
MPS that assists the customer to optimise their fleet, meaning a
reduction in the sales volumes of hardware units.

This, in effect, adds mounting pressure for the office printing
and document imaging industry to navigate a sound path forward.
The perfect storm – as many people call it – has only prepared the
industry to accept a more challenging and constantly changing
future that involves not only the physical printed page, but also
a future in how digital and paper-based information and communication is delivered and managed across a customer's business.

It is important to note that over the last three or four years, if
you had asked either myself or the select few, highly-regarded,
global, independent thought leaders in the office printing and
document imaging industry – such as Greg Walters, founder and
President of Walters & Shutwell and the president of the MPSA,
or Ed Crowley, CEO of Photizo Group – we may have predicted a
different outcome from what we see today.

Only a few years ago before the start of this decade the industry seemed reluctant and disinterested to acknowledge the need

and the speed of change required. If some had acknowledged this at the time I would expect it would have certainly been done at a much slower pace. In my view we have been fortunate as an industry, with companies and organisations such as Photizo Group and the MPSA continuing to promote the value and become the voice of MPS to the industry.

By bringing together relevance and a new value proposition through a more effective business model, MPS became the perfect vehicle for the industry to leverage from and to get behind. It has become the beacon that has allowed them to transition their business from a product-led sales model to a services-led business model.

Up until then I expect some manufacturers were oblivious to the required changes that needed to take place both inside their own businesses and, more importantly, how they addressed their position of relevance in the eyes of their customers. Managed print services have become the impetus for the industry to change.

Today we see the industry has quickly adapted its business model to become more services focused. In fact, up until only a few years ago, (prior to 2010), the industry was not even allocating or recording a profit or loss under an MPS ledger. It was still product segment (for example, Segment 4 relates to 41 to 69 pages per minute) and unit-number related.

Although there are a number of industry gurus who deserve special acknowledgement, I have highlighted these two gentlemen (Walters and Crowley) because they are both knowledgeable and insightful on what is taking place within the industry and because they are beacons as they continue to strive, support and promote this always-exciting industry. At the same time they have challenged it to continually evolve and become an even better, stronger, industry for the future.

For me this is part of the reason I have published this book. I feel propelled to not just sit on the sidelines and make frivolous comments and judgement calls. As a player with a long involvement with this industry it was my duty to assist and help develop a platform for other industry leaders and experts to build off my comments or challenge my point of view, both in the Australian marketplace and across the world. As I learned long ago, there is more than one truth in the world. My view and comments are from my set of lenses.

I know myself, Messrs Crowley and Walters and a number of my peers around the world continue to put themselves in the firing line for two reasons: the first is that we are doing it for and on behalf of the end customers; and the second is for all the actors and players that operate within this fascinatingly simple, yet complex market sector. We all work hard to help the industry progress to the next stage of evolution.

In Australia I expect to be the brunt of the tall poppy syndrome that means, for anyone outside Australia, everyone (specifically my peers) will want to cut me down, bring me back to earth. It's supposedly the Australian way so someone does not get too big for their boots.

Even though I have been cautiously reluctant to write this book knowing the brutality of the tall poppy syndrome, it also rallied me to take this step and journey regardless. Hopefully my shoulders can carry the load and lift up others to go further.

CHAPTER 11:
MANAGED PRINT SERVICES JUST BECAME LEGITIMATE

A SERVICES FUTURE

Even with this perfect storm the industry appeared to be gaining a new lease on life. We certainly started to see increasing merger and acquisition activity across the industry after 2010. This may have been evolving through a growing acknowledgement that MPS was gaining legitimacy. A managed print service was now both timely and relevant as a services offering. Many customers had firsthand experience of the financial and operational effectiveness of a managed services contract.

A services maturity had now been reached and MPS made sense to the customer especially to the areas of finance, IT and strategic sourcing and procurement. The office printing and document imaging industry saw MPS as their way to transition to a new business model and capability. A business capability based on a new approach, a services-led model, rather than their traditional hardware-led transactional sales model.

Managed print service was also pertinent to the organisation as more and more office printing devices were being connected to a customer's network. IT promptly assumed accountability for managing and responding to these assets from a technical standpoint.

Even though terms such as fleet management, device management and global services programs may have existed for over a decade, many vendors had their own unique terminology or program offerings which made it difficult for the end customer to determine and understand the differences between the offerings and the value they provided.

Also customers prior to this period, had still seen these program offerings very much as a single branded offering. Many customers were either not ready or were still reluctant to turn over their fleet, or parts of their business, to a single brand for all their office printing, document imaging, digital workflow and process improvement, mail room, and print rooms.

Fuji Xerox Australia's Global Services division may have been the exception in the early period with a number of managed services clients. However they were still challenged to convince the overall market that an all-encompassing offering was the best.

MPS as a more recognisable, quantifiable and consistent services product really received a bounce of momentum through the Photizo Group who in 2009, led the charge in this space and established what we now know as MPS. Later that year the MPSA was officially founded.

It is important to acknowledge that "the MPSA is a global, non-profit organization that provides independent communications, collaboration, education, standards and success to MPS professionals. The MPSA is not owned or directed by any one organization; it is in fact owned by its members. The influence of any single company, large or small, is prevented by the MPSA charter".

Following the establishment of the MPSA a more overarching

definition was derived to meet a wider range of multiple audiences. The definition of "managed print services" as defined by its member is:

"Managed print services is the active management and optimisation of document output devices and related business processes"

THE ONLY GAME IN TOWN

As a go-to-market model, MPS was quickly adopted by virtually all office printing and document imaging OEM brands around the world. (This included both A3 device centric OEMs as well as the A4 device centric OEMs.) Through this momentum it also prompted resellers, dealerships, agents and hybrid sales organisations to become more independent in what they offered to the market.

A number of brand-dominated dealerships quickly realised that being brand-independent had its advantages over its disadvantages. The noose and control of the large OEM brand had loosened and control and flexibility transferred back to the independent solution providers but more importantly back to the end customer.

Many IT resellers, dealers and hybrid solution providers saw being independent was perfect for their engagement model to the customer. It also provided them the flexibility to capitalise on the A4 product devices coming through the indirect distribution channel of HP, Lexmark, Kyocera, Samsung, Oki, Brother and others. By having the variety of these A4 devices and brands and in combination with the A3 OEM brands, the resellers had the capability to provide a solution for every business requirement in regards to a hardware solution fit.

Overnight the resellers shifted from being brand-led to solution-led. With this capability, MPS rapidly took off both in Australia and around the world from 2010 onwards.

In Australia, before this time, there had been a slight moving of deck chairs as different dealerships switched camps from one brand to another. However this was rarer in the 1990s and early to mid-2000s, mainly due to the control that OEMs of A3 centric device brands had over access to their products, spare parts and toner consumables.

On the other hand, A4 centric devices that evolved from the office printer were, in most cases, distributed through an indirect channel so control was more with the product reseller than the manufacturing brand. Every dealership or independent had access to an A4 centric printer brand available for either hardware sales or service support, depending on their service capability.

The rise of MPS in Australia allowed independent solution providers, traditional dealerships and resellers to insist on access to both types of product lines from more than one brand. To provide MPS, providers had to be able to provide A3 led multifunctional devices from the traditional manufacturers such as Canon, Xerox, Ricoh, KonicaMinolta, Toshiba, Lanier, and Sharp; and A4 office printers from HP, Lexmark, Kyocera, Oki, Samsung, and Brother.

At the same time we began to see the ever expanding range of multifunction printers from both the A3 manufacturers and A4 led OEM brands being distributed and sold through both direct and indirect distribution channels.

In Australia most office automation equipment and IT products are sold through dealerships and IT resellers and distributors if they are not sold by an OEM's direct sales force. In the past, in Australia, the office automation industry used dealerships to build market share and coverage for a particular A3 OEM brand. That is one reason why dealerships are still very prominent in Australian regional locations.

Although some office equipment dealers in metropolitan locations successfully competed head on with the larger A3 OEM direct sales force, this was more the exception than the rule in the decades prior to 2010. However we are now seeing more metropolitan based dealers position themselves, with multiple brands or as independents, directly competing against large direct sales teams of the OEMs. Their success is focused at the lower end of the customer segment: small and medium business customers where profitability is usually high and the dealers are not in direct competition with OEM direct sales. These dealers do, however, compete against sales agents, resellers and the OEM's driven dealers.

Another reason why dealerships are more common in regional areas, today, than metropolitan locations is that the OEM's direct business model is more costly to sustain in a regional location. By implementing dealership programs it provided both the A3 and A4 centric original equipment manufacturers the opportunity to reduce the cost of sales, extend its geographic reach, reduce its direct risk and increase its opportunity to increase market share and unit volume.

The A3 branded OEM devices certainly needed a strong dealership program in Australia in the past. A strong dealership network was essential to win and support large direct enterprise customers that required a national footprint for service coverage and service response times to their offices in regional or remote locations.

In Australia, with approximately the same landmass as the United States but a population around 14 times smaller than the USA (22 million) direct OEM operations were not financially sustainable outside most of the capital cities.

Today, in Australia, many of the A4 based engines from both A3 and A4 OEMs are pushed through IT distributors and resellers. IT distribution is a more centralised affair with most major IT distribution companies operating large distribution centres in major cities or states and pushing IT equipment and office products through a range of dealers, resellers and IT service providers to service both regional and the central business district or metropolitan locations.

THE INDEPENDENT PROVIDERS

By early to mid-2000 we saw a few independent or hybrid print solution providers who operated as part sales, part contract broker and brand reseller, move into the market. In this role they would work with the end customer to identify their office printing costs and review their pricing contracts. They would advise them on current market pricing rates to establish cost savings, negotiate with the brand and then sign a contract with the end client. In Australia this has been relatively successful due to the customer's feeling that some OEM brands displayed a level of perceived arrogance and dominance over brand and contracts in their dealing with them.

The customers felt that having someone who was from the industry and acting on their behalf would achieve a better outcome. On many occasions this worked and the customer gained substantial cost savings. In some respect it bought the customer to the table with the incumbent OEM brand and forced them to discuss their grievances and concerns. In these situations the independent solution provider acted as the mediator to move the process forward.

Some large A3 OEM brands actively worked against these independent solution providers because they feared the risk of losing control over their client's business or the client realising what their real costs were. For others it was about seeing their profits reduce.

On the other side of the fence, other A3 OEM brands stayed impartial and open to the process. They saw the risk of losing an alternate channel to market as too great while, just as importantly, gaining access to a new client that they may not have had access to before was too good to pass up.

In the past the office printing and document imaging industry in Australia was effective and responsive to a customer facing dissent although, with this new change, the industry took longer than it should have to re-establish its customer focus. Nonetheless the A3 OEMs have now started to control, or more effectively limit, the independent solution hybrid providers by not providing them access to their A3 devices. At the same time a number of the A3 OEM device providers have become more willing, intent and aligned to the end customer's requests and this has reduced the end customer moving away from their large A3 OEM incumbent.

Regardless of the situation some A3 device OEMs will not engage with independent providers even if they run the risk of losing a potential client opportunity; even if the end client continues to feel aggravated with their incumbent OEM or the customer feels that the process of dealing with multiple OEMs and resellers or dealers is not in their best interest. It is clear that a number of OEMs today are prepared to take this risk on as they firmly believe that this position of dealing directly with the OEM is a far better outcome in the end for both them and the customer.

This change in posture by the A3 OEM may have a major impact on some of the independent hybrid providers as their A3 business model relies on clipping the ticket on both the capital and service rates that were provided to the end customer. In their model they would effectively re-invoice the client with a different capital price for the equipment and service rate for the contract term. The independent provider would receive the margin difference between the OEM hardware and service price at and the price that the independent provider quoted the end client. In effect it is similar to a wholesale rate and a retail rate.

By the way, this is no different to how A4 devices (single functional printers or A4 multifunctional printers) are sold. Although the outcome is the same, how you package it can be different; for either A4 devices or the A3 devices from OEMs. The difference between the two is that the OEM manufacturer pricing is rarely transparent to the market.

BEWARE THE PRICING OF INGREDIENTS

Although I am no chef, or even an authority on how to cook, the simplest explanation I can come up with is this. The ingredients that you have to bake a cake may be the same, but imagine if what you baked actually tastes different to the last time you baked it, or does not taste how you expected it. How could this happen? Well maybe you added the ingredients in a different order, changed the volume of an ingredient or had to substitute something. The way you add, bundle (mix) or charge these combinations of ingredients into the mixing bowl can change the outcome of what is delivered.

This is the same for many A4 centric branded devices such as HP. A4 centric devices have more broken down ingredient options so each customer may receive a different end product. This makes it very difficult to compare price effectively, or somewhat convenient, depending on which side of the negotiating table you are sitting at. When mixed in with an A3 device, or a mixed fleet of A3 or A4 devices, then the solution takes on multiple dimensions.

Even if we go back and consider what the independent provider is offering, we still see that they have a role to play in the market. A number of end clients are very happy with this arrangement and believe that the independent provider provides them with a single point of contact and contract management and that the arrangement they have is far better then what they would have or usually receive from an original equipment manufacturer. Other customers may not be fully aware of, or do not see the benefits of, this contract practice at all.

MARKET MATURITY

One major outcome transpired in Australia through the combination of the independent solution providers' successful competition against the direct OEMs and the growth of MPS from 2009 onwards.

What occurred was an explosion and fragmentation in the number of independent providers, dealerships and office printing sales agents, a multitude of hardware and software service providers and integrated hybrid solution providers, new market entrants from sectors such as IT service firms, IT resellers, toner consumable supplies businesses, office products suppliers and the new type of hybrid independent MPS firms all starting to market a MPS offering to the market.

As these businesses grew, the direct original equipment manufacturers had to cut through and rebuild their channels to market. Effectively they had more competition. To counter this they had to review the channel programs they offered. As an outcome we see an increase in partner programs that support more agents, more metropolitan and regional based dealers who may now provide an assortment of brands, and a variety of value-added resellers and solution partners that go to market from non-traditional office printing and document imaging segments. In short, the end customer is being rapidly educated by more suppliers in the market.

Through this process we now see an array of providers who expect to deliver a better customer experience. They are aiming to provide a more cooperative, integrated and fit-for-purpose service to the end customer. They are outcome-focused, not product-focused. These providers, as a minimum, provide a base line of both an A3 device portfolio from an assortment of A3 branded OEMs and A4 OEM branded devices from the printer centric OEMs. I still refer to hardware being the base line as most providers are still renumerated on sales revenue and generation of income and profits through device deployment and the more valuable toner consumables.

However, today it is at the next level above product deployment where providers will differ in their next level of offerings. We are seeing the industry quickly divest as they engage and manage their client's business.

At this period the more successful providers appear to be shaped around a small number of key customers rather than a mass-market approach. Their intimacy with their customer's operation becomes their defining point of success and through the intellectual property of their client's business or through their vertical industry segmentation, understanding and willingness to stretch to their customer's requirements, they can and have been able to adapt more quickly to their client's changing business requirements.

This relationship engagement, improved customer focus and overall management acceptance is occurring across large enterprise and mid-market businesses as well as small businesses in both corporate and government entities.

The continual and ongoing marketing efforts of both local and global businesses promoting MPS have increased. Clients have become more aware of the business benefits that MPS offers. Through this process, it is fair to say, the level of customer maturity has increased to a point where sales and marketing firms of MPS now have to create a stronger market offering and differentiation between what they offer to market and what the competitors are offering. The difference this time is that their competitors don't look, talk or act the same as they did in the past. The evolution continues.

CHAPTER 12:
TO PRINT OR NOT TO PRINT? THAT IS THE QUESTION

GIVE ME AN IPHONE OR TABLET ANY DAY

This chapter highlights a very quick synopsis of the changing trends of office users while illustrating the potential and future impact of the printed page. Office printing is declining and will continue to decline in the future. The key question is at what speed will this occur and what are the influences that are affecting this transition?

However be very clear. We will not be entering a paperless office for a number of reasons – well at least not for the next 10 to 15 years.

OFFICE PRINTING IN DECLINE

We have all heard about the paperless office, it has been consistently mentioned over the last 50 years. It was first mentioned in mid-1964 as a publicist slogan when the IBM 2260 was launched. Later it was mentioned in 1975 when Business Week published an article "The Office of the Future" predicting that office automation would make paper redundant for routine tasks such as record-keeping and bookkeeping and it came to further prominence with the introduction of the personal computer.

Although the worldwide use of office paper more than doubled from 1980 to 2000, it has been a little harder to track throughout the last decade or so as worldwide print volume statistics appeared inconclusive. For example, Lyra Research (now Photizo Group) estimated that 15.2 trillion pages were printed worldwide in 2006 while IDC suggested that, between 2007 and 2010, more than 10 trillion pages were printed in offices in the U.S. alone.[7]

More recently we have seen the use of office paper levelling off and, in 2012, saw office printing decline for the first time, according to Photizo Group[8] and IDC[9].

There are a number of reasons why office printing volumes are decreasing worldwide.

INCREASED PROCESS AUTOMATION

More and more back office processes are becoming automated as paper-based forms and documents become digitised in areas such as human resources where leave and vacation applications and sick leave forms are now mostly completed online.

We see more content and documents sitting in document management and content management systems rather than being always physically available. As new documents or content is produced it can be automatically updated in multiple business systems or integrated into other business processes electronically.

Additionally we see process improvement increasing the efficiency of paper-based systems with batch scanning and batch processing in departments such as accounts payable and account receivables allowing invoices to be electronically managed, stored and processed. This reduces the need to print pages as part of the approval and authorisation process. We also see exception reporting for faster processing of digitally scanned paper-based documents, such as authorisation signatures and account information details like home loan and credit card information.

There has also been a much stronger business focus on increasing overall business efficiency, reducing operating costs and improving customer satisfaction levels. This is being achieved through implementing and integrating business wide operational strategies, practices and disciplines such as business process re-engineering, business process improvement, business process management and business process outsourcing.

USER AND BEHAVIOURAL CHANGE

It is reported that the user need and requirement for office printing is believed to be falling. This, in some part, may be due to a generational shift in our working population. The Millennials, Nexters or Gen Y – born between 1981 and 1998 – are the digital generation. They are said to be less inclined to print and are more comfortable with, and used to, receiving and reading information on a screen.

Although this may be the case there is also compelling evidence reported in journals such as the Scientific American. A recent article titled "Why The Brain Prefers Paper" suggests "people often understand and remember text on paper better than on a screen. Screens may inhibit comprehension by preventing people from intuitively navigating and mentally mapping long texts. In general, screens are also more cognitively and physically taxing than paper".

Going on from this point there is a strong view from a plethora of experts and academics that one of the reasons why the US college system has seen student's academic scores declining recently is that students may not able to manage and learn effectively due to the vast amount of digital content they are exposing themselves to.[10]

Although they can access a great deal of information the unfortunate side effect is that they are not digesting much of what they read or learn. Human beings may not be able to effectively interpret and synthesize the amount of source or content volume of information. Many students who are connected to the internet are continuously scanning information.

Today college students are using a wide range of digital and social media to stay connected. They use applications that drive and source information from the internet. Content from applications and products such as Facebook, Twitter, Snapchat, Instagram, iPads, iPhones, Kindles and the like become endless.

In fact some would espouse that they are the poster boys and girls for the consumerisation of IT. More on that later in this chapter.

Therefore the challenge becomes how to manage consumption and retain valuable content. It is continually suggested that scanning or surfing the web trying to absorb large amounts of information from a screen may be impacting student retention levels and impacting their results at exam time.

As they navigate across the web they are becoming seduced. The learning becomes easily interrupted by other applications with pop-up messaging. This distraction adds to what I call surface learning. They understand the high level message but do not have the commitment or focus to continue to read the whole content, book, case study, article and the like due to a multitude of competing online distractions. Therefore it may be at exam time when they are required to provide more information they may be lacking the ability and depth to explain the content and leanings in more detail. Time will tell, if paper-based learning is better.

But they (Gen Y) are not the only ones who are experiencing this problem of distraction. We see this today in the business world with applications such as email. Today office workers become obsessed with their email inbox, baby boomers included. Their productivity can be impacted by email and uncontrolled access to social media applications at work. Even in business we have learned to skim read to be efficient, we even communicate (talk, email and text) in short, bite-sized chunks. Maybe we are suffering the same symptoms. How do you retain information better: from a screen or a printed page?

Although it has been argued that Gen Y's entry into the office will have a major impact to office printing, I think there is enough evidence and speculation to challenge this view. In a business context there is certainly a requirement for specific information to be used and shared through the use of the physical printed page. I know not everything needs to be produced on paper. But this does not mean we will be a paperless office society in the next one or two decades.

THE AGE OF MOBILITY PRINTING

The increasing use of mobile workers and mobility allowing workers to work from a variety of locations such as hot spots, coffee shops, airports and home has increased significantly over the last few years and will continue to increase.

There had been a strong industry view that if a business

implemented mobile printing, office print volumes would decline. Although there may be some truth in this, it is still too early to confirm the severity of the impact. I expect there may be a drop off but, potentially, only a slight variation as I expect that office users may in fact print the same as they did in the past. They may just print to a different location than the business expects.

Although users may be no longer be stuck behind their PCs if they desire to print something, the ease of printing from a tablet or phone may not change the desire to have something printed out on a page for the same reasons they may have printed the document if they were still behind their PC. The real issue may be where that printed page will be printed out.

We will see printed pages being printed at locations outside the usual office environment. This in turn may add a future hidden cost to organisations that may not be tracing or accounting this type of expense today. In this scenario I expect that the cost of a printed page may be more expensive externally than the costs of their current office device. Something the business will have to watch out for.

It is also unlikely that the increasing usage of mobile printing will be able to completely offset the ever-reducing single functional desktop printing volumes in the office. The real story is that the print volumes from single functional devices are simply being redistributed to higher volume workgroup printers and multifunctional printers. I expect much of the mobile printing volume will go to the same location (centralised workgroup printers and multifunctional devices) rather than smaller single function printers in the office.

MOBILITY AND LACK OF STANDARDISATION

One of the real disappointments with mobile printing is that the industry continues to shoot itself in the foot regarding standardisation. There is no simplification or standardisation when it comes to developing a single platform to resolve the complexity of the ecosystem for the benefit of all users. Recently a new consortium, the Morpria Alliance, has brought together a number of brands – founding members Canon, HP, Samsung and Xerox – to form a group standard. However, we still continue to see a wide surplus of varying device types, file formats, operating systems and platforms from most other brands.

One reason for the lack of standardisation is that the OEMs may see mobile printing as a threat to their most important profit engine: toner consumables revenue.

The issue is twofold.

First, mobile printing allows multiple users to print to their choice of branded device and, at the same time, it takes print volumes away from other competing device brands.

Second, office workers may choose not to print on their local office device at all, and send print volumes to a competing device (outside the office environment or another branded device on another floor or location), therefore impacting toner consumables revenue for whoever manages their local printer.

BYOD AND THE CONSUMERISATION OF IT

A newer trend that is impacting the end customer environment is the consumerisation of IT and "bring your own device" (BYOD). BYOD is supposedly more strongly driven by the younger, tech-savvy, Gen-Y workers, however I think this is now relevant for all generations. They prefer having their own devices and applications rather than receiving company mandated devices and being limited to accessing only employer-approved applications.

From a business perspective the real impact of BYOD with regard to office printing is the management of access to a printing device's hard-drive and the level of security from these personally owned devices hitting the network.

Combined with mobility printing, a locally saved file, or document, that is printed at a remote print location – say the local coffee shop – will need careful consideration in terms of what is being printed. Securing client information to protect confidentiality with regard to what may be printed will come under increasing focus for both government and commercial industries. Imagine your private health or finance details being printed out at the local coffee shop through the use of mobile printing by someone who had access to your information from inside the company that you entrusted with your details.

GREEN IT

In Australia we have seen a strong emphasis for large enterprise organisations to shift to a more socially conscious and green IT culture. As part of the organisation's cultural transition it has been common for large enterprise organisations to move into new energy efficient premises. In doing so the enterprise attempts to cement new business, operational and social practices. At the same time the business hopes to gain the additional financial benefits that come with such a transition.

Through this redesign and a growing focus on activity-based workplaces we see office staff moving into locations that provide a hot-desking environment. Hot-desking effectively eliminates an office worker having the same regular desk each day. In a hot-desking environment you arrive each morning and sit at any desk you want.

What this also means is a change in work practices such as where employees store their information, documentation and files that used to be in their office draws or filing cabinets. These new work areas have virtually eliminated personal storage areas and filing cabinets that took up tremendous floor space and were an expensive and unnecessary cost in floor real estate.

Eliminating this physical element has meant a more prudent electronic printing and filing practice is required by users within the business. As a consequence, users are more considerate of what needs to be printed. Many organisations have been able to encourage the "print less" philosophy by designing and implementing electronic filing systems and processes that make it easier for users to manage this transition effectively.

SCAN RATHER THAN PRINT

Since mid-1996 we have seen the marketing term "distribute and print". This was derived from the capability of digital or network printing where we could send electronic files (documents) from one location to another via the network (via email for example) and then print to the nearest output device.

Today in 2014 we are seeing a practice of "scan or upload and publish". More and more businesses, big and small, are shifting to multifunctional devices. Their decision to acquire this device is not only for its printing capabilities. The ability to scan hardcopy documents to the network is becoming more important to businesses that want to keep their knowledge accessible and available to everyone at any time.

We also see changes in how original content is being created, managed and shared. Businesses and their office workers are producing content, then uploading it to a document management system, web content management system or the cloud, to publish it to the web, to nominated recipients or to make it available to all parties.

IDENTIFYING WHICH PAPER-BASED PROCESS TO AUTOMATE

Best Practice

Although many businesses would prefer to back-scan all their old and existing paper-based hardcopy information, we recommend that you do not do this in the first instance. This can become very expensive to implement and manage, in a number of ways. In many cases we expect that the costs will outweigh the benefits.

In our experience, clients should take a different approach. We recommend that the business audit a number of current paper-based processes to identify the five most valuable internal and the five most valuable external documents. There are a number of ways to consider what is valuable. The value will depend on the client's individual business driver, for example improved customer satisfaction, increased responsiveness, increased productivity, reduced human labour or short and long-term cost savings. These should be discussed more fully with business owners and key stakeholders to substantiate the value contribution.

Identifying the most valuable pieces of paper helps to understand which process should and could be automated. It will also indicate which areas should not be considered in the first phase. As this process may run across the business, the early identification will also identify the right key stakeholders to manage and be involved in the process. It will also help to identify and budget what technology may be required. In many cases a plan such as this can gain momentum by providing quick wins for the program initiative prior to taking on extensive projects.

We also recommend clients benchmark their competitors when it comes to paper-based or electronic processes in order to identify how their competitors manage the same challenges.

It's important to map the processes and workflow carefully without assuming that manual and repetitive processes are good places to start. Although these areas may be touched as part of the outcome, it is sometimes more effective to not look at current processes. Always keep in mind the business objectives that you are trying to achieve. Existing processes may be a distraction from gaining quick wins and meeting financial and business objectives.

Based on the plans and objectives that an organisation may choose, we suggest that businesses start with a project to transition a select group or area of paper-based documents into an integrated and automated workflow process, or that they simply start to scan only new hardcopy documents that come into the building from selected departments. Either of these two options can be very effective.

We also recommend not making these changes throughout the business until a number of control projects have been implemented to measure further improvement opportunities. Measuring before and after is crucial for the project's success.

Section 4:

TONER IS THE REAL BATTLEFIELD

CHAPTER 13:
MORE THAN THE BOX

The real Trojan horse of the office printing and document imaging industry is not the physical devices they install in your office, it is the toner or ink that puts marks on a page. The original equipment manufacturers (OEMs) may provide the vehicle that connects to the client's network and transports the document through the device, but toner is the fuel that delivers marks on a page. It is this fuel the OEMs now rely on to provide them with their profitability.

From this aspect, the game is really on. The players are fighting to wrest control and win market share in an industry that has potentially reached its peak (in office printing) and is now facing a future of successive declines. The first decrease showed itself in 2012 and this may not be the last. Office print volumes may continue to decline year on year. International Data Corporation (IDC) research shows that "worldwide, page volume from digital hardcopy devices decreased to 2.98 trillion in 2012 from 3.03 trillion in 2011, a decline of 1.5 per cent year on year".[11]

This fact continues to raise the question: is the office printing market shrinking? If it is, how will the industry and the players react to an ocean that is starting to shrink and dry up? This challenge may drive some OEMs to take risks to ensure they can preserve their income while building for tomorrow.

In covering this, what are the tell-tale signs that print volumes may continue to fall and what of their impact to toner supplies, the lifeblood and profit engine of the OEMs?

Also are there any lessons from related industries, which have also found themselves in shrinking markets in recent times? Is the office printing and document imaging industry just the next natural industry along an evolutionary curve to be impacted? Moreover could the office printing and document imaging industry be on the same trajectory and does it face a similar fate?

Many of these industries have been impacted by technological change and innovation, an increase in new market entrants, competitors that have greater agility through less operational legacy, innovative business models, a transition in consumer behaviour and increasing global competition. All these mean the industry has to adapt much faster than in previous years and decades.

Specific industries that have felt this pressure more recently include the photographic film industry, the publishing industry, a variety of bricks and mortar retail stores who did not see online retail as a competitive threat. To review these questions we will consider one particular industry that is somewhat correlated with the office printing and document imaging industry. We will look at the personal computer industry, tablets and PCs.

The relevance of the tablet and PC market is interesting for a number of reasons. One reason is the speed of change in user technology adoption. The other is that both the PC and tablet directly impact on printing devices, print volumes and toner supplies.

Tablets, as a category, have exploded at the expense of the personal computer (PC) market. Latest reports from Gartner (reported through www.techcrunch.com on October 21, 2013) suggest worldwide tablet shipments to grow 53.4 per cent this year alone. The total number of shipments in 2013 was around 184 million units. While traditional PCs are still shipping a lot more units (303,100 million forecast for this this year)[12], those shipments are continuing to decline, and are predicted to be down 11.2 per cent on 2012 shipments.

In a relatively short space of time we are seeing the tablet steal significant market share from the PC market.

On the office printing and document imaging side, figures from the Photizo Group suggest a seven per cent decline in compound annual growth rates for hardware (office printers and multifunctional devices) and a 0.8 per cent decline in growth rate for supplies between 2013 and 2017.[13]

This data set could provide an interesting link to understand if there really is a correlation between the PC industry and the office printing and document imaging industry. Will one industry influence and impact the other?

To examine this point further we must understand the difference between PCs and tablets. As a very simple characterisation the PC generally comes with a full screen, a keyboard and mouse. One of its core appeals is its ability to design and produce content. On the other hand, tablets, to continue the simple characterisation, are portable and generally used to consume media and view content.

With this view a tablet is somewhat passive in creating content. Although this is changing through the variety of platforms and applications available, I just wanted to keep the two products separate for the sake of this comparison. I expect as the tablet evolves it may have a larger impact on the PC due to its increasing functionality through increased and improved applications.

From this simple characterisation, we can start to raise and breakdown a number of relationships that may exist between a PC and your office printer. Firstly there is a relationship between the PC and printer in the context that content is created using a

PC to be printed on the printer. Therefore content creation and content output could be considered more traditionally coupled and reliant upon each other as part of a business and user process in respect to a normal office environment.

Printing, therefore, has been somewhat reliant on a user having a PC to create content such as a document. As users or knowledge workers became heavy content creators over the last two decades print volumes increased dramatically. Has the transition from PCs to tablets eroded content creation and reduced print output within the office? At present it is too early to tell if and what the preceding declines in PC sales are having on office printing.

Although mobility printing utilising a tablet (or smart phone) may rebalance the print volume equation, it is still important to note that the relationship exists. Personally my view is that we are still in a market transition mode at present and it is too early to tell, but I have a sense that if the PC market continues to decline this will have a larger impact on traditional office printing and the subsequent print volumes.

TONER: THE MONEY GAME

Manufacturers have long been giving up the profitability on the sales of hardware devices as the game was not about the printer or multifunctional devices being delivered and installed at the customer's site. It was all about the toner that ran through the device, providing income and profits for all those who touched it on the way through. In some way it was money for jam; each copy, each print, each click was cents and dollars to the supplier. Now the manufacturers will have to reinvent themselves as printing output starts to decline within the typical office setting.

Sure the industry has been able to show resolve and bring new products to market that kept its evolution going. This time around it may be a bit more difficult to sustain its income from the output of hardcopy documents the way it had before. The multifunctional device as an information hub within the corporate or office environment has not eventuated like the automatic teller machine (ATM) that they have hoped for. Well, not in its present form.

Personally I think the industry will continue to sustain itself going forward through two specific strategic decisions and actions. The first is to continue with retooling and leveraging their operating platform, the applications that exploit embedded software, communication and information services that support the business and users. I expect the device will play a more pivotal role in interconnecting to business applications, other devices and mobile user communications as systems and technology communicate with each other in the near future.

The second is to diversify more heavily into providing clients a range of business and IT services that support and extend value to the customer's operational infrastructure, a service not necessarily related to a device.

COLOUR KICKER

On reflection, colour printing didn't have the impact it should have or the industry expected. Although colour printing was a fabulous income-generating vehicle for a relative short period between the late 1990s and early 2000s, competition between the brands drew pricing discounts to the table faster than the industry would have liked or been prepared for.

From an industry perspective, a number of office manufacturers at the time may have been forced to play their cards earlier to either gain market share or were forced to reduce their service rates on colour just too compete. Service consumables pricing for colour output dropped from an average of 35 to 40 cents down to around 12 to 15 cents and now it sits at a commercial rate anywhere between 6 to 7.5 cents. Today colour still represents a cost factor of 10 times the cost for black and white output.

Obviously, the continuing technology research and development at both the device and toner manufacturing level over this period advanced. This may have allowed the industry to re-price their colour consumables model. This would have had the desired effect, at the time, of increasing volume penetration into the market as colour toner pricing moved into a more affordable price range.

Today when analysing a customer's fleet of devices we continue to see a fairly consistent ratio of 75 black and white prints to 25 colour. Back in the late 1990s we expected an even split (50:50) between the two based on the expectation that colour was a more effective and powerful way to communicate a message and drive customer action. Those ratios never really materialised.

Now I know some clients will not have these black and white to colour ratios as colour printing may be much lower or non-existent. Some specific vertical markets and business will not require colour printing at all, so this ratio will not fit everyone. But as general rule I expect to see clients who have colour-enabled devices as part of their fleet to be around this ratio. Let's just say it's a good rule of thumb. Or, if you particularly like the Pareto rule of 80:20, you could use that and not be too far wrong.

PRICING TONER

With this said my question is what price does colour printing need to drop to in order to increase the ratio of colour printing to black and white? Although colour printing is enormously profitable to the manufacturer, especially in comparison to black and white printing, will lowering colour print costs trigger customer volumes to grow? If this did occur, would some manufacturers be able to transition customers away from the incumbents by undercutting price again? Reducing colour print costs for current customers is not a strategy I see any manufacturer wanting to play unless it is a strategic way to win new business.

At this juncture I'd like to forecast five things that I expect will happen next, in relation to colour printing within the office environment:

1. Firstly, I expect to see price reductions again soon as I believe that an industry brand manufacturer will feel the need to go after increased market share. It will only take one brand manufacturer to become nervous or desperate to drop their pricing. Even if this act is for a limited time, it will drag other brand manufacturers into a retaliatory position, which extends the pricing discount period.

2. The second will be the advent of the new inkjet technology from firms such as HP, Brother and Memjet that will challenge the elasticity of colour pricing here in Australia and around the world. This new development in inkjet quality (page wide array) and cost will put pressure on current pricing and this will push colour pricing lower, even if only used as leverage from the customer position.[14]

3. Continued advancements in software technology will help to improve the costs of colour by effectively using less colour toner when printing. For example, software such as Adobe LeanPrint is designed to

reformat the images using software algorithms to remove unnecessary background images and display ads on web pages.

4. I expect that colour printing costs still have some room to move (decrease) as part of a wider strategy when implementing a MPS contract. Colour pricing discounts may play a larger role is delivering a more cost effective offering to clients. As part of the ingredients mix of a MPS contract, colour may be the trade off when trying to lock-in a client for the next four or five years. Although colour print reductions may not be as noticeable on the surface to a client it will provide the essence to change the pricing structures for the client's contract. The challenge for the industry will be to keep this pricing strategy quiet. Now that's virtually impossible in this industry. As soon as one salesperson decides to jump ship and move to another supplier the game will be up and everyone will be trying it. It will be like a virus that you can't stop. The OEMs know it so they are more reticent in running these pricing programs. However this all plays well for the client to reap the rewards.

5. The final component that will pressure and squeeze the colour print pricing is related to two conditions that we are seeing more frequently as part of the offering of a MPS contract. That is increased flexibility and low-cost options for purchasing toner cartridges.

It is important, at this point, to restate that a MPS model is not based on the specific hardware, colour, style, shape, size or appearance of the device, its logo or the software that will need to be installed on or embedded in the device.

The MPS model is based on the business outcome the client expects to be delivered. From this aspect many MPS organisations may attempt to standardise the product portfolio range. At the same time they may attempt to consolidate and rationalise the number of devices in the existing fleet and so on.

However saying this we are seeing two elements or conditions starting to take shape with many MPS being delivered to the client. I believe this will continue to happen as part of the process of becoming more efficient in delivering a cost effective MPS solution to the client.

The first aspect is around the product set that is being delivered. There is an increasing use of A4-based engines which are cartridge-based devices and form part of the larger product portfolio mix delivered to clients where footprint, device flexibility, and user volume or cost ratio are all important.

It is also important to restate that customers, and specifically a client's procurement or category division, are more acutely aware of the increased sales and marketing activity from office supplies industry giants in Australia such as Staples, OfficeMax, Officeworks, Lyreco and Complete Office Supplies and the extensive online toner suppliers who promote continually lower priced cartridges. These cartridges are either for premium toners or lower cost compatibles toners. These two conditions increase the concentration on price competition to lower colour printing costs.

We will continue to see greater competition and lower costs come from internet-based toner supplies companies selling toner at a margin of 20 per cent or less, against the 30-50 per cent gross profit margin of the bricks and mortar suppliers like Staples.

COMPOUNDING DECLINES

It's not all doom and gloom for the industry just yet. They are not leaving without a whimper. I know the industry will continue to protect itself in the same fashion it has always done, by focussing on increasing revenue attachment rates in the typical areas that surround the box such as finance, professional services and software and an array of other business services.

But as we are seeing, device unit numbers are reducing. On July 30, 2013, Photizo Group reported that "worldwide unit shipments will fall to just under 88 million units in 2017, down from 111 million units in 2012".[15]

This may be, in some way, due to the increasing success of implementing MPS. In the majority of cases the aim of a MPS provider should be to optimise your existing fleet. Effectively this may equate to right-sizing the fleet both in terms of either print volume or the number of devices. This, in turn, may mean reducing device unit numbers.

Beyond this initial approach MPS is really about assisting the client to gain increased operating efficiency as well as a continuous improvement in managing the associated print devices, printed output and workflow integration of both hardcopy and softcopy documents.

As device numbers decrease, attachment rates will decline in unison as they are linked with device installations. This will have a compounding effect on the industry's major players. Potentially only small indicators at first but, as competition hots up, we may see a frenzy as some bottom feeders fight to gain market share.

Even though it may be at a loss, two, three, four or five years of locked-in toner consumables supplies sounds pretty good. I say pretty good as this may be the difference between being in business or being out of business.

This may be easier said than done. It is always a very difficult task to change your business model and approach when you continue to make healthy or reasonable profits globally. Should they now start to sacrifice profits as they turn to a new future, which may in the initial period, offer less profitable returns? Is there a way to transition without harming the existing business? At what speed does the transition occur? Is it easier to stay the course and be more aggressive on winning more of the market share from your existing competitors?

PRODUCTION PRINT

One parallel area that needs to be brought into focus is the high volume production printing and publishing devices sold to large corporate businesses such as law firms. They typically take up large floor spaces and their centralised locations usually have names like "the print room". Outside the corporate office you would see them in instant print shops and commercial printing companies. This device is the bigger brother to the typical office printer you see in the office today. Office devices and production devices are somewhat more challenging to delineate due to a number of variables, such as print speeds, print volume capacity, application requirements (for example, variable and production publishing) and the level or need for a dedicated operator to manage print runs.

For this book I have kept production devices out of scope with respect to the general office printer and multifunctional devices that I'm discussing. However there is an aspect that I would like to make an association with.

The connection point is that colour printing costs on these production devices still sits at or around the same print charge as the office devices. Some people would argue, quite rightly, that they are different engines that produce different output quality, dare I say it, more like commercial offset print quality – especially when colour reproduction is involved. As a veteran of production printing I'm the first one to be very clear on the differences. At this point I am not challenging anybody's view on the differences except for one lone point.

That is, over the last 25 years in Australia, production printing (or coping) output or print charges or click charge rates have, for the most part, been lower in price point or click charge than an A3 engine office multifunctional device. Today I'd argue that this is not the case for colour. This then raises other questions about the base point of colour print charges today in the market.

Are manufacturers creating or holding a pricing floor for colour print charges? Are they controlling the pricing of colour prints across both category segments (office and production)? Or are colour printing costs at their lowest possible manufactured price?

CHAPTER 14:
MANAGED PRINT SERVICES ARE NOT A SILVER BULLET

MANAGED PRINT SERVICES: IT DEPENDS

The question that I am continually asked by end customers, IT resellers and dealers both in Australia or when I speak at conferences overseas is, "What are managed print services really?" Everyone appears to be looking for that "silver bullet" answer: a word or two, or sentence that says it all.

Some salespeople and some customers expect that there should be a line item called MPS in the price book. Some offer, and talk about, MPS as a product-led and product-only based offering to customers. I know this because I have seen it many times in the last four or so years.

But saying this I also fully appreciate why this question is being posed. When a business or a person has relied so heavily on buying or selling devices (or boxes) the process becomes very distinct, defined and identifiable as a tangible product offering.

We now expect these same end customers, as well as many players from within the industry, to understand how to buy and sell a services-led offering and deliver on that capability through a process of consulting. The business opportunity is no longer as clearly defined around a tangible and consistent product offering as it was in the past when there was only hardware and some software to sell. The overall business problem or business solution may be far broader than the previous hardware based discussion path.

As experience and maturity sets in over time, I expect the process of consulting and analysing will become more methodical and definable as part of a services-led offering. In the very near future the scope of the problem or solution will include a set of tangible products including hardware, software applications and a variety of services in a box (pardon the pun), such as document workflow and imaging and content conversion. They may also include ongoing management of the end-to-end capability and ownership of the contract delivery, print or otherwise.

More importantly MPS gives sales companies the opportunity to have a wider and deeper conversation about the customer's business. If you are marketing and selling MPS to your customers, expect your business to change. If you are already successful in MPS then you're probably already seeing your business stretching or transitioning to something new.

MANAGED PRINT SERVICES SHOULD BE SIMPLER, RIGHT?

I'm an avid supporter of many people's view that we should try to make things simpler for all parties involved, especially the end customer. But that is always a little bit more challenging to achieve when an industry such as the OEMs has instilled in its workforce a way of selling and marketing its offering. This business success or business model of the past and present may today be creating a legacy as it attempts to move forward.

In fact, the threat and impact of how the industry has operated both locally and across the globe, especially around the growth opportunity of MPS, would certainly be perplexing to the industry OEMs. Strangely enough, MPS is, in many ways, somewhat opposed to the OEM's previous manufacturing based operational model.

Alternately, part of the reason for the complexity may be based around a fear that a push towards simplification could drive the OEM's businesses, or the industry, closer to becoming commoditised, or MPS may be complex simply because it is different or new to many.

Managed print service is still evolving. It is, I believe, now starting to move into a new level of maturity, however that maturity is still only MPS version 1.0. The adoption rates of MPS are increasing. Managed print services have moved from customers at the enterprise to large businesses and it attention is now focusing on small to medium business that are now seen as more profitable.

SO WHAT ARE MANAGED PRINT SERVICES?

I will share some answers and definitions of what an MPS is, or is becoming, later in this chapter. As I have stated, there isn't a one line item answer. Managed print services may be relatively new to some, but to many others it has already become mature in its offering.

To the purist, a MPS is a philosophy or process of consulting with the customer to identify and understand their print and document imaging needs. It's the opportunity to engage, analyse and identify the larger business problems and opportunities around the process of moving both hardcopy and electronic documents around the organisation.

MPS as services provision provides the chance to be more tightly integrated and be more highly valued in your customer's business, rather than delivering a transaction-led hardware product.

For the non-purist it could be simply providing toner supplies or paper for devices when required, or servicing and repairing a desktop printer the next day. This too could be highly valued by the customer. It's not for me to decide or define; it's the customer.

Therefore, MPS means a lot of different things to different businesses and individuals. It may depend, somewhat, on where an individual or business sits on the "engagement curve" (the intersection of where the sales or business conversation is taking place) through either its products or the services that it offers to its end customers.

In my view, the true managed print service offering is much wider than just the physical device and printed output. Managed print services can include everything pertaining to both the physical device as well as how documents or content in a physical or digital format is converted, transported, outputted, reproduced or even processed within businesses.

A managed print service has quickly morphed into a wider and more integrated sphere of managing digital content. Worldwide the office printing and document imaging industry is now transitioning to match the new business services opportunity. They are setting up, deploying, investing and acquiring businesses that provide them increased capability to control and manage both physical and digital information and data. This includes the capability to own or manage the processes and technology that underpins it. Their future services business will now be shaped and focused on how digital information is shared, repurposed, moved and managed around their client's business.

HOW BIG ARE MANAGED PRINT SERVICES IN AUSTRALIA IN TERMS OF REVENUE?

Before we look at Australia, let's see what is happening around our region and the world from a number of industry analysts' reports on MPS and basic print sales numbers.

What will become clear is that each analyst measures the industry differently. This again highlights that there is a level of complexity in measuring this industry. However, all seem to agree that MPS will continue to grow from now up to 2017.

Although MPS appears to be growing, I would suggest that two key elements need to be considered when seeing the revenue numbers the analyst firms have put out.

The first point is that it is only in more recent times (over the last three to four years) that the industry, including some of the leading OEM organisations, have transitioned their revenues and financial accounting from previously independent hardware unit sales, service and maintenance contracts and software profit and loss into a new MPS profit and loss. Therefore, as more firms move to centralise their financial accounts into a MPS profit and loss, I expect that worldwide MPS revenue will increase through a level of MPS revenue consolidation.

The second point is in respect to how MPS is increasingly capturing previously uncontracted pages (generally toner-based cartridge purchased ad-hock, such as for an A4 laser printer). As more pages are captured under a MPS arrangement (contracted pages) I expect to naturally see an increase in MPS revenues.

I'm not of the opinion that MPS is increasing because customers are buying more devices or printing more. My view is that MPS growth is coming via global profit and loss consolidation and existing uncontracted pages shifting to contracted pages under a MPS contract arrangement.

With this said, let's look at worldwide numbers and how they drill down to Asia Pacific then into Australia and New Zealand MPS revenue numbers.

Globally

Let's look at the worldwide market. We see that MPS is expected to continue on its growth trajectory. According to InfoBlog's Randy Dazo, MPS will grow on 6.7 per cent compound annual growth rate from 2012 to 2017 and continue to increase from approximately USD$20 billion to USD$28 billion dollars over that period (see Figure 1).

WORLDWIDE MANAGED PRINT SERVICES FORECAST 2012–2017 IN MILLIONS OF DOLLARS

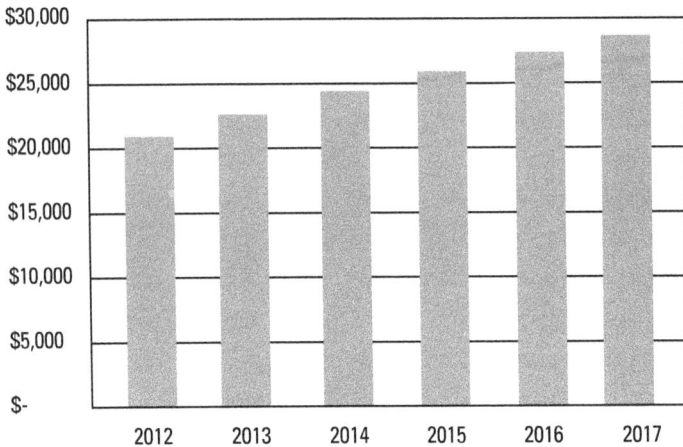

Figure 1: Worldwide managed print services forecast, InfoTrends

Photizo Group also forecasts similar growth rates, forecasting the total worldwide MPS market in 2014 to be over USD$21 billion and by 2017, more than USD$27 billion. Still at a macro level, the combined worldwide MPS and basic print services market, according to IDC's Marketscape (worldwide managed print and document services 2013 hardcopy vendor analysis), suggest growth will increase from USD$25.1 billion in 2011 to USD$27.9 billion in 2012 with a year-over-year growth rate of 11.1 per cent. IDC predicts that worldwide MPS growth will grow from USD$10.8 billion to USD$11.8 billion in revenues at an 8.9 per cent compound annual growth rate (CAGR) over the same period.[16]

TechNavio's analyst forecast published on October 2013 suggests that global MPS market will grow at a CAGR of 10.75 per cent over the period 2012-2016.[17]

As you can see there are slightly different growth comparisons in terms of growth forecast and time periods. However the key take-away is that all continue to predict growth will continue until, at least 2016.

Regionally

Looking across the region, IDC reported through Business Wire in January 2014 that revenues in the Asia/Pacific (excluding Japan) (APeJ) print services market, which includes MPS and basic print services, will cross the US$6 billion revenue barrier by 2017. IDC also says 2014 will be a pivotal year for market expansion into broader customer segments as well as further market growth.[18]

Photizo Group are predicating the Asia Pacific region, including Japan, will hit around USD$2.5 billion revenues in 2014 pushing out to over USD$3.1 billion in revenue by 2017.[19]

Australia and New Zealand

Closer to home, the Photizo Group forecast that a combined Australian and New Zealand market for the MPS industry will hit just over the USD$880 million dollar mark in revenues for 2014 and that by 2017 the market revenue will increase to around USD$1.0 billion dollars. InfoTrends, on the other hand, suggests that the Australian MPS market will be closer to USD$613 million by 2017.[20]

Today the Australian market is dominated by the OEMs. The top three players, Fuji Xerox, HP and Ricoh, are the most dominant OEM brands today across both direct and indirect channels.

So, for this region and specifically Australia and New Zealand, the market opportunity looks like it will continue to grow in terms of MPS revenue.

WHY MANAGED PRINT SERVICES ARE NOT ALWAYS THE SAME

Why aren't MPS always the same? This is a good debate that continues within the office printing and document imaging industry. For some it may be sometimes complex to explain but there is no shortage of passion from anyone involved in this exciting and ever-changing industry. From my perspective there is no wrong answer. I will add, though, that passion does not mean that someone is always right – well at least not all the time.

My personal view is the somewhat annoying cliché, "it depends". In my experience there is a lot of truth in the phrase "it depends". When I say that, I am basing my view on the situation at hand. There are sometimes many factors and variables to consider on any particular given situation, both now and into the future, that it's difficult to provide a 100 per cent guaranteed outcome. Sure you can build in a disclaimer to protect your recommendations or the solution you offered, but this can have a level of complexity in itself. Therefore, in my view, you have to take each situation and customer challenge on its own merits and understand that the variables can change.

The good news is a solution will have a level of base construction, key elements that need to be applied and integrated as part of the ingredients framework. But a MPS solution will almost always need to be customised and aligned to the customer's situation and requirements.

Today, most good service providers should have a framework or methodology in building out a consistent solution for the customer. I expect some are efficient at building repeatable processes with consistent service deliverables. We know repeatable scenarios do exist, but you have to be careful not to go in with too many pre-existing thoughts, ideas or solutions. This is one aspect that still continues to see services providers trip up. Service providers have a tendency to go in with the attitude, "it looks like, it smells like, it sounds like and it feels like our standard MPS offering". Well this approach will hurt the service provider and more importantly the customer.

Packaging up MPS may be a quick and easy approach to get the sale. But this approach, done incorrectly, falls straight back into the old transitional sales mode. MPS, if not managed pragmatically and attentively will come back and bite – hard.

This chapter, and book, does not seek to provide all the situational analysis and variables for every occasion for each customer objective and goal required. It's really more about both the "supplier and customer beware".

However I will illustrate, in this chapter, how all potential players (suppliers) along the office printing and document imaging engagement curve may have a role and function to play in the eyes of the end customer. It may look insignificant in our eyes but it may not look that way to the end client.

Here are two relatively extreme examples to prove the point. In one case we have someone selling consumables (toner) only to an end client. They promote and market this service as a MPS offering. In this scenario I am not going to take them to task by saying they are wrong and they should not be using MPS as the vehicle to sell toner. Sure it may not be right from a purist view of MPS and I accept that.

As an alternate example, should an office printing and document imaging company helping an end client integrate and distribute information on to their web content management system, or helping them improve the customer's document workflow, be considered more of a pure play managed print service business? Whose decision is it? You, me, the big OEM or the customer?

Nevertheless that's exactly the point I am trying to get across to our industry: we have to be open enough in our minds to know that this may be where the offering may start and evolve. We cannot condemn an opposition play where it connects on the engagement curve to a client's business, or where a document or image may be shaped, formed, materialised in the physical world or shared within the digital world in our client's business.

There are many parts and relationships that make up our industry and all are, and can be, as important as others. To suggest one is more or less important than another will blind your vision of the future. In many situations you'll see that it depends on the customer to make what they think and believe is the right decision for their business at the time. This does not mean that you cannot provide leadership direction or a recommendation to the customer, as the customer is looking for this too.

MANAGED PRINT SERVICE IS A SERVICES PROVISION

To provide an insight into why there has been an increased adoption and acceptance from both the customer side and the sales organisations for MPS, you have to understand how the industry is transitioning from a product-led to services-led approach and why customers have an increased appetite for leveraging services rather than just buying product hardware.

One of the main reasons is that today's managed print service offering is not shaped around a brand or necessarily an explicit hardware or product item. You are selling an outcome, a service delivery, a provision of service or a continuum of service. Your service may include a multitude of combined and disparate components such as device hardware, onsite service, valet service, fleet monitoring and supplies management, depending on the client.

Customers ultimately want to pay for an outcome. Effectively, if the service providers do what they say they will do, when they will do it and deliver in a timely, cost effective and value-contributing way then you, as the customer, do not really care how big the devices are, how many there are, what type of device is on the floor or the desk, what colour the buttons and devices are, which brand label or logo they have on them or what the service engineer looks like on a particular day.

All you, as a customer, really care about is the solution you are paying for. Is it fit-for-purpose? Are the services you receive positively contributing to your business? Will it provide the outcome that you have requested and approved? Is the service in line with what you are being asked to pay? You would expect that, as a minimum, it does what you have asked for. Ideally it provides more than you expect and exceeds customer expectation.

My favourite way to explain how a provision of service should work is based on the following example of hiring a taxi.

When you call a taxi cab to go to the airport, do you stop at the curb when they arrive and refuse to enter the vehicle and not take the ride because he or she arrived in the wrong brand of car?

No.

What you may care about is whether the taxi is clean, if your luggage fits, the driver's knowledge of the best way to the airport at this time of day. Will they get me there safely and on-time? Will the fare be reasonable? Will they have the right music for my journey? Will the ride be peaceful so I can relax and think? Will the ambience be right? Is there air conditioning? Please don't let it smell bad!

Obviously all these are important to you, the customer and yes, unfortunately, in most cases we would still take the trip to the airport without all of these criteria being met. However, we have not focussed on brand and, in a managed-services world, we would probably expect all criteria to be met.

In a managed services world the customer is certainly paying for an outcome and requires a number of key criteria be met, or exceeded, by the service provider.

Now a test: can you recall the brand of car that took you on your last three taxi trips?

My point is that customers are moving away from the product elements such as the brand of printer or MFD and now acquiring the services outcome. More specifically, the brand of the device is becoming less important when the client is buying a services provision.

Clients do not see MPS as disparate products, where the brand of those products becomes a key criterion for decision-making. Obviously there will always be elements of the service that are important and may play a more pivotal part in the services capability on offer. However with many of the devices and brands today, differentiation is becoming harder, though not impossible, to define.

The most important consideration is still the execution of the MPS. It is not always the ingredients that make it a success. You could have the best products but the whole services provision or customer experience may not be at the level of an alternate services provider with inferior product but a higher level of service.

CHAPTER 15:
TAKING THE PRINT OUT OF MANAGED (PRINT) SERVICES

As the name suggests MPS is a managed service and, for the most part, many from our industry would expect MPS to be pertinent when the sales engagement, transaction or service provision is directly attached to the physical devices or the print output specifically.

But, as we are now starting to see, the future will not always be about the physical printed page. It could be considered, therefore, that managed PRINT services is deceiving from this aspect as it's suggesting you manage PRINT only – the physical page.

That raises the question of what a managed print service is. I will provide more detail in the next chapter to help you develop a more informed view of what MPS is or could be. But all definitions continually suggest MPS is about more than just print.

For this reason, I want to now focus on the broader concept of managed services. Firstly the managed services model is not new and the level of maturity of what it is, how it can be shaped and scoped, how it should be managed and what to expect of a services provision is, from a customer's standpoint, not new. The customer is far more experienced in this space.

Based on this understanding, I would like to make the point that, firstly, our industry is new to this game (managed services), and our customers are not. Yes we are transitioning fast, but the industry could be considered somewhat untested, potentially inexperienced and, to some extent, unproven. Therefore perception would see us trailing in this new business services capability model of managed services.

Secondly we need to always understand the process is not always about boiling the ocean. Some customers prefer a process of testing, incubation and controlled growth. MPS like most managed services may start as a single point solution in the beginning then expand their scope outward over time.

Most medium and large enterprise customers would have at least one form of managed services contract today and many of these would have started from this same starting point.

For example, a managed services contract could be as simple as providing a waste paper pick-up service to ensure secure and compliant disposal of information. This may then expand to managing an organisation's entire internal office printing, the scanning of hardcopy originals into digital or electronic workflow systems for the client's customer relationship management system or managing their accounts payable system and processes.

Most medium to large enterprises are choosing to outsource non-core business functions, processes or departments like human resource administration, help desk and call centres, data centre management, cleaning and maintenance, office supplies, voice and data communications, IT infrastructure and so on. These strategies are usually driven to increase operational efficiencies, improve customer satisfaction and reduce cost.

Holistically, most managed services offerings have continued to evolve and change. That said, managed print services should be seen as a process of evolution that integrates the supplier's business capability with the client's business needs and requirements, starting with office printing and document imaging as a beachhead. This could be transacted at any stage as a functional requirement, an operational or technical or a service-led capability. It can be more about how you establish and support the customer in the initial stages as well as how you and your customer adapt and transition with each other over time.

Reiterating, MPS may start with device management and may, over time, progress into additional services that are based around what the provider can deliver today or what the client would like it to deliver tomorrow. The reason for this unique change in posture is that, over time, you should have gained a position of trusted partner rather than just a single point service provider. Valued strategic partners have the opportunity to become more integrated into their customer's business, therefore increasing the value of the relationship and services provided.

So in short, don't define your business by the products or services you sell, but more the contribution and value you provide the customer.

MANAGED SERVICES AND MANAGED SERVICE PROVIDERS

Before we provide definitions on MPS, I think it is helpful to consider definitions from managed services perspective without the word print included. A number of definitions that try to define managed services extend to defining a managed service provider. I could continue to expend definitions on all the relevant IT terms but it would not be helpful or productive.

The challenge is that there are a number of definitions depending on which technology, industry or product vertical you are focused on. In other words, technology products across tele-communications, data and software each have their own industry perspective on managed services.

To keep this topic straightforward I have used two references, Gartner and Wikipedia, which may not be the most appropriate source but I include it because that definition is becoming more accepted.

Wikipedia says:

> *A managed service is the practice of outsourcing day-to-day management responsibilities as a strategic method for improving operations and cutting expenses*

It goes on to say, "this can include outsourcing HR activities, Production Support and lifecycle build/maintenance activities. The person or organisation that owns or has direct oversight of the organisation or system being managed is referred to as the offerer, client or customer. The person or organisation that accepts and provides the managed service is regarded as the service provider or (MSP)"

Based on this definition, I would define a managed services provider as:

> *A managed services provider (MSP) is, typically, an information technology (IT) services provider that manages and assumes responsibility for providing a defined set of services to its clients either proactively or as it determines that the services are needed*

Wikipedia also tries to capture a wider level of service scope for MSP with its inclusion of the following statement "most MSPs bill an upfront setup or transition and an ongoing flat or near-fixed monthly fee, which benefits their clients by providing them with predictable IT support costs. Managed Service Providers (MSPs) are sometimes contracted to manage multiple staffing vendors and to measure their effectiveness in filling positions according to a customer's standards and requirements.

In effect, the MSP serves as a neutral party that offers the customer a complete workforce solution while ensuring efficient operation and leveraging multiple staffing companies to obtain competitive rates. MSPs typically use a vendor management system (VMS) as a software tool to provide transparency and efficiency — along with detailed metrics to the user — related to every aspect of the contingent and contract workforce. The model has proven its usefulness in the private sector, notably among Fortune 500 companies, and is poised to become more common in the government arena. MSPs often use specialised software, in order to control and deploy managed services to their customers known as Remote Monitoring and Management (RMM) software".

I have included the MSP definition and how MSPs function as the similarities to office printing and document imaging providers will have them asking, "what type of business are we really?"

Gartner defines a managed service provider (MSP) as:

A managed service provider (MSP) delivers network, application, system and e-management services across a network to multiple enterprises, using a "pay as you go" pricing model. A "pure play" MSP focuses on management services as its core offering. In addition, the MSP market includes offerings from other providers — including application service providers (ASPs), Web hosting companies and network service providers (NSPs) — that supplement their traditional offerings with management services.

As you can see Gartner's definition takes on a broader application focus for managed service.

The managed services definition from Wikipedia is less specific in the technology but clearly defines the provision of service offered and the objective of the service.

In summary the managed services definition and the managed service provider definitions illustrate that there is no wrong definition, just elements of variability and level of experience depending on the business and your individual set of lenses.

More importantly the key take away is that all definitions provided could be overlaid when it comes to MPS. A managed print service just happens to be more defined due to the word print.

CHAPTER 16:
MANAGED PRINT SERVICE DEFINITIONS AND MPS 2.0

I have regurgitated the definitions at the end of the last chapter to provide further insight into how MPS was shaped and formed from an industry and supplier need to fit a changing customer's business strategy and operational model and also how it needed to align to a customer's maturity in managing varying non-core service provisions.

The following definitions of MPS provide further insight into what many believe is MPS in its simplest form. As documented in Chapter 11, the definition comes from the MPSA, the only international, independent and non-profit MPS organization that embraces all industry participants in a collaborative environment.

Wikipedia also uses the MPSA definition for MPS.

MPSA says:

Managed print services is the active management and optimization of document output devices and related business processes

The second definition comes from InfoTrends, a leading world-wide market research and strategic consulting firm for the digital imaging and document solutions industry.

InfoTrends says:

"Managed Print Services are a services-led offering that can help companies solve their pain points typically around the management, cost and/or document processes, by delivering continuous improvements, particularly around an organisations' document and output environments"

I believe these are the two that neatly describe MPS. I recommend that both end customers, and the industry as a whole, use these two definitions as a starting point.

Two more definitions describe MPS. Both are below. Both demonstrate that there is no single approach to the problem.

Quocirca is a research and analysis company primarily focused on the European market. Much of Quocirca's work is sponsored by a broad spectrum of IT vendor, service providers and channel organisations. Quocirca says:

a "Managed Print Service" is the use of an external provider to assess, optimise and continuously manage an organisation's document output environment in order to lower costs and improve productivity and efficiency while reducing risk.

From Gartner, a world leading information technology research and advisory company, says:

"Managed print services (MPS) are services offered by an external provider to optimize or manage a company's document output".

Gartner goes on to explain, "The main components provided are needs assessment, selective or general replacement of hardware, and the service, parts and supplies needed to operate the new and/or existing hardware (including existing third-party equipment if this is required by the customer). The provider also tracks how the printer, fax, copier and MFP fleet is being used, the problems, and the user's satisfaction".

The key take-away from these definitions is that MPS continues to be a moving target and the current definitions try to capture a wider footprint

THE CUSTOMER WANTS A NEW RELATIONSHIP

In a product-led sales and marketing model (transactional sales process) the desired outcome was usually the transaction itself. The buyer acquires the goods from the seller and then uses the product. An income stream may be attached to this sales process, derived from ongoing service and maintenance to maintain the product, but the focus of the sales organisation is the pre-sales activity, the processes between and the outcome of a sale. In most cases, once the sale has been achieved the sales organisation moves to the next customer opportunity and starts the process again. It is repeatable for each and every client.

This has been a very efficient and profitable process for the sales organisations in our industry. It was the way the sales organisations built their business models for the last 30-plus years. Over that period the end customer has been valued in this process in most instances. So it's now a difficult concept for many salesperson and sales organisations in our industry to simply walk away from, or at least acknowledge that something that has always worked is no longer necessarily right for the future.

In most cases, the end client now wants more than just the old transactional relationship. They still may buy in a transactional way but they expect a business relationship to be part of each and every transaction – even if the transaction is a simple, price pointed, engagement. The customer still expects that this relationship will hold many of the same characteristics as their more complex, strategic and critical investments. In either type of investment situation the client expects the relationship is built on trust, integrity, dependability and it inspires value for the client. I guess this has never changed, but it certainly now appears more evident in the ongoing engagement.

This fact continues to be observed in our own customer engagements when acting as business advisors to finance and IT. It is also stated in our customer feedback interviews. Clients are increasingly annoyed when sales organisations don't align their sales engagement with the customer's needs and desires. One particular area of noise surrounds business development managers and account manager roles.

As the office printing and document imaging industries evolve they will need to be more mindful of managing the process of first engagement through to ongoing management of the customer relationship. This will be more challenging, as the cost of sales is more difficult to budget and forecast when the sales activity becomes dormant on a particular customer account.

Due to this difficulty, some sales organisations may still require their sales model to provide both the old product-led transactional engagement and new style of services-led sales. Although in some cases, at certain times, this may be achievable and appropriate, the real challenge is to know how and when the customer is transitioning in their approach to the relationship status.

So as an industry it is certainly a balance that many industry players are trying to get right. However the most challenging aspect is the balance of managing the return on investment and opportunity costs from the side of the sales organisations and while meeting the demands and expectations of how the customer wants and expects to be dealt with.

MPS 2.0?

MPS 2.0 will be the evolution of services from what you see now to those provided by the application capabilities that sit in or on the platform (operating system) of these multifunctional devices. As I said on my blog (www.first-rock.com) "Move over MPS we are moving to The Internet of Things (IoT)".

The office printing and document imaging industry will transform in the coming years through its ability to adapt and advance its existing business and technology frameworks. The days of an office device just printing, copying, scanning and faxing will be the thing of the past. These intelligent information hubs will drive more interaction between themselves and other devices.

In a recent article in ITwire, Graeme Philipson said, "...that The IoT encompasses hardware (the things themselves), embedded software, communications services and information services associated with the things". He went on say, "that Gartner refers to the companies that provide the hardware, software and services as IoT suppliers. The incremental IoT supplier revenue contribution from IoT in 2020 is estimated at US$309 billion."[21]

He continued, "due to the low cost of adding IoT capability to consumer products, Gartner expects that 'ghost' devices with unused connectivity will be common. This will be a combination of products that have the capability built in but require software to activate it and products with IoT functionality that customers do not actively leverage."

Your little devices that are sitting on your desk now, or out in the corridor, may just become a lot more interesting. Could they be the next ATM in the office? Could they wirelessly power your mobile phone and tablets? Will they be able to scan your body and measure your blood pressure? Will they be able to turn on or off, report and monitor your other IT products?

Sure there is always an ebb and flow debate between centralisation and decentralisation and this is certainly valid in respect to computer power (for example storage and memory) and where applications should reside. However, the increasing user pursuit, and business requirement, to manage a never ending barrage of user-driven technology and tools brings into focus technical trends such as "bring your own device", mobility services, cloud applications and the consumerisation of IT. Businesses will need to consider how they best support and deploy these areas of technology in the future.

We are already seeing applications on your mobile phone carried across to your printing and MFD devices, making the device just like any other computer that interconnects and manages data today. In business, this may become even more prevalent as it starts to drive new business services.

Maybe it will be called "the IT network in a box".

This is where MPS 2.0 will come into its own, the evolution will continue.

Section 5:

THE RISE OF
MANAGED
SERVICES

CHAPTER 17:
NEW MANAGED SERVICES – RED OR BLUE OCEAN?

I've chosen "red and blue oceans" to help me provide a framework and relationship to what I see is happening in the Australian marketplace. This chapter outlines how the office printing and document imaging industry, in its many forms, is feverishly and continuously repositioning itself in the Australian marketplace. As an onlooker, commentator and end customer advisor it's interesting to see how each player is trying to gain traction in areas outside their previous domain of device hardware sales and generally single-pointed software sales.

I've used red or blue oceans to leverage the discussions using some of the principles of Chan and Mauborgne's *Blue Ocean Strategy*[22] that discusses a concept of blue and red oceans. I see the similarities and comparisons between what they illustrate in their book and how the industry, both globally and in Australia, is trying to adapt.

In particular I see the original equipment manufacturers (OEM) continually challenging themselves to transition into new industry sectors and markets with a combination of old and new business models at their disposal. A great industry that is trying to re-invent itself quickly and one we certainly wouldn't want to see shrink in any major way. But therein lies the real challenge for the industry in Australia: can it transition successfully and sustain itself or will the market decline at such a rate that some major players will be forced to exit; or will industry players consolidate or merge to keep a foothold in the market?

SO WHAT IS THIS BLUE OCEAN STRATEGY ALL ABOUT?

In short the authors of *Blue Ocean Strategy* believe that an organisation should create new demand in an uncontested market space, or a "blue ocean", rather than compete head-to-head with other suppliers in an existing industry – in this case a "red ocean".

Blue ocean

The blue ocean is all industries that do not exist today. It's the unknown market space. Competition, as yet, have not realised it even exists. In a blue ocean, demand is created rather than fought over. Blue oceans have abundant opportunity to grow profitably and usually this occurs in rapid fashion. In a blue ocean competition is irrelevant not only because it may not exist, but more commonly because the rules of the game are yet to be set. In short, blue ocean is an analogy to describe the wider, deeper potential of market space that is not yet explored.

Red ocean

A red ocean represents all the industries that exist now. Companies within the red ocean try to outperform their rivals by obtaining a greater share of product or service demand. As the market space becomes more crowded, the prospects for increased profits and growth are diminished. In a red ocean, products and service offerings become more commoditised or niche and therefore competition becomes more cutthroat, which turns the ocean bloody – hence the term "red ocean".

This chapter aims to bring both a perspective and insight of the industry players and how some of these industry players may be simulating or adopting strategies that may be considered a blue or red ocean play. For some it may be just about swimming, staying afloat and surviving. Others may be moving less consciously, only realising their mistake as the frog did when sitting in the boiling water and when it's far too late to do something about it. Others may be already circling their next prey.

IS IT THE SURVIVAL OF THE FITTEST AND THE MOST ADAPTIVE?

More importantly, behind this chapter is my real deep interest in how the Australian industry players are pursuing their strategy transition. My concerns are derived from my observations, conversations, questions and insight about where the industry is headed globally as well as here in Australia. Is the industry in Australia doing enough to develop blue oceans? Do the players have the autonomy to make the change into blue oceans locally? Can they sustain their business while making the transition? Do we have the market capacity to establish new blue oceans?

The good news is that I know a handful of leaders in Australia who are probing at some form of blue ocean. I know many will see this as a too risky play at this point in time, so they will be forced to follow their overseas parent's strategic long-term plan. Unfortunately, many will discover that what they think is a new blue ocean is really a red ocean in disguise. These so-called camouflaged oceans will put them under even more pressure. Potentially they will have to battle across a number of fronts as they realise that they have actually entered someone else's ocean after all.

Most may be more comfortable, and conditioned, to sit back and wait for a new blue ocean to appear through someone else's development. Others will jump into another red ocean in an attempt to try to demonstrate to the market that they have a strategy of some sort in order to be seen to do more than nothing. Some will try to cover all their bases and actually operate in a number of (red) oceans at the same time while making a play into a new blue ocean, like a gambler covering their bets.

In my view the OEMs and the wider industry face a continuing two-pronged challenge.

1. First the OEMs and all the market participants in our industry have built an income and profit model that has been very successful over the last 25 years. That can be a hard legacy to overturn. Today it's a shrinking market. Both revenues and profits year on year are declining both globally and in Australia, with very few exceptions. Although some quarters that provide hope, and areas such as consumables are increasing thanks to growth in bulk toner shipments for colour and black and white printers. These good stories may only be short lived (next three to four years) or appear more as an accounting irregularity that provides good news stories going forward.

   Additionally, industry OEMs have continued their long-term investment in research and development, product innovation, sales, marketing and distribution through direct subsidies or through indirect channels. These cannot be shut down, retooled or diverted quickly without a major impact to the current business and operational model.

The parent operations, as well as their subsidiaries around the globe, face the same challenges at the same time although, some may argue, the new economies may preserve some players for a little longer. However, I am not certain this will provide the kicker required. I believe the new economies will adopt newer technologies faster than most expect. Why would they want to adopt the past when they have the opportunity to leap frog the front runners? These new economies are building new infrastructure to support both their internal population growth and their international competitive advantage. The facts are there: the markets, or oceans of the past, are increasingly red.

2. The industry's second challenge is to change the customer's perception of the brand itself or the OEM industry as a whole. The problem starts and ends with an existing legacy, its brand or logo.

The OEM industry has developed powerfully recognisable brands and logos that are symbolic and trusted by many of us. But today the customer's perception is that these powerful brands are still very product (hardware) focused and potentially similar to consumer-level products. Changing this perception to a managed services business will be a challenge in itself. Some players find it harder than others to transition the customer's perception.

There is no doubt that the brands are very well regarded in all aspects of technology and price performance. However this has created a legacy that is difficult to overcome in the eyes of the customer and decision makers, especially at finance, IT and procurement levels.

Many of these brands have shifted into what I call "consumer-class identity crisis". What I mean is that the brands are so well known, have been around for such a long time and most have either a red or a blue logo. They have become very comfortable to both business and consumers. This level of comfort may be influencing business buyers in a negative way. They may see these brands as consumer, or commoditised, technology products rather than business solution providers or consultancy and application integrators.

Client feedback would suggest that some brands are finding it more difficult to be taken seriously in new endeavours such as managed services. Following IBM's successful move from products to services over a decade ago it has been somewhat more difficult than expected for the office printing industry to make this same transition. It would appear the customer, especially from an IT perspective, is not as willing to make this leap of perception. However saying this there are always exceptions and the industry is used to being in a fight.

3. Increasingly the challenge is that clients want to make the transition away from being hamstrung by print output based technologies. This will certainly open up a new level of challenges, opportunities and new conversations.

SWIMMING IN A RED OCEAN

If you are in the office printing and document imaging industry in Australia today, you are certainly seeing an ocean that is turning a shade of red. Now, no matter what everyone is telling you or who is telling you, the signs are there and they have been building for quite some time.

With approximately 22 million people, the Australian market is relatively small in comparison to many western economies. However, the competition within our market is incredibly aggressive. We have an abundant number of OEM brands, direct, indirect and online sales companies, reseller and IT channel partners, office supplies companies, dealerships and sales agents as well as IT managed services and IT outsourcers all targeting the same end clients at multiple market segments with an endless supply of printers and multifunctional devices.

This goes some way to explain why Australia is considered mature in areas such as MPS. We have had to become agile and more resourceful when adopting a MPS play as one of the key challenges that impacts MPS execution is distance, time and logistics.

With the same land capacity as the USA our ability to economically and efficiently manage a client's dispersed fleet across multiple locations throughout both metropolitan and regional Australia becomes even more difficult with mixed fleets. Our ability to sell and market MPS had to be innovative from the outset.

Also, the way the industry was set up and cut up in major capital cities by the host of sales companies is somewhat scary to behold. The predatory sales activity across states, cities and postcodes predetermined much of the aggressive sales competition we see today between salesperson within their own team and company, let alone against other brands and resellers.

WHAT'S TURNING THE OCEANS RED?

Some pockets or currents of light blue oceans are starting to appear but they are very shallow indeed. In fact, an overnight storm may just as easily wash them away. However, that does not mean progress should be abandoned, not at all. It may be the opposite: the industry may need more prudent investment if it is to fight back the rising tide. There is an undercurrent of change and all the players are looking for that next big wave to ride. Some, I expect, will wipe-out along the way and one or two may even drown.

As the key players look to transition they will need to overcome many of the key business challenges that all players within the industry are continuing to face:

- A market that is shrinking globally (office printing volumes are shrinking)
- Revenues and profits that are being pegged back or in a state of decline
- Increasing competition for local market share
- Managed print service may be having an adverse effect on the devices and consumable sales in terms of unit and print volumes:
  - As MPS gains more success its ability to optimise an existing customer's fleet will reduce device unit numbers per client which reduces both print volumes and consumable supplies.
- Reductions in retained earnings reduces reconciliation income remitted to parent operations:
  - This may have an adverse impact on future product pricing for new equipment parts and consumable supplies coming into the country which will either drive local pricing higher and reduce market share, or

- The OEM could reduce local operational costs to maintain its pricing and use this lower pricing strategy to push for future market share.

• Many of the direct operations carry a cost base that is too high for the size of the market although there has been some recent cost shredding activity by one or two firms.

• Reward and recognition plans are no longer sufficient for a supposedly services-based business and sales behaviour is still driven by device volume unit activity.

• New suppliers continue to enter the market from China, Korea and the USA

• New, so called, disruptive technology platforms such as page wide-array which boasts ink-jet colour printing at virtually half the cost of current laser engine based toner devices.

• Shrinking markets and industries that have an impact on office printing industry, for example, the decline in the PC market as businesses and users switch to tablets and paper manufacturers closing plants around the world.

TURNING THE OCEAN BLUE

To avoid red oceans or, as some would say, to resist the virtual rabbit in the headlights position, industry players including OEMs may have to foster a new approach to innovation. Although there have been systematic strategies like Porter's Five Forces Model[23] for operating in red oceans, there is no such formulated strategic methodology for creating blue oceans.

To create a blue ocean strategy, businesses from within our industry may have to decouple their existing premise of moving, or bringing existing current clients who they work with today, through to new market opportunities. This approach may appear to be at loggerheads with much of today's traditional thinking.

It may be more advantageous to go after a new set of clients, or customers, in the new chosen markets (blue oceans) as opposed to staying on familiar ground where organisations have been successful before. It's certainly moving outside your comfort zone to decide not to take your existing customers into new markets. Finding new customers for new markets may be both more innovative and also preferable for a number of strategic reasons.

The danger in taking your existing customers into blue oceans is that the product you are offering may not solve the needs of your existing customers. It's not that there is a problem with the offering, just that it suits a customer that you haven't met yet.

Both perspectives have to be tried. New offerings can appeal more to clients that you have never worked with before than to existing clients due to perception or an existing legacy with your existing relationship.

The challenge, if presented with poor results from your existing customers, is to look beyond your customer segment to all the other aspects of your business: the products, the sales and marketing, the solution pricing. Everyone will try to convince you that your customers love you (which they might) but that may not be the love you need for your new blue ocean market. You may need to find a new love. If you don't find new customers you will continue to hunt in the red ocean.

The challenge in adopting a blue ocean strategy is breaking out of a red ocean by making the competition irrelevant as you enter uncontested markets. If you want to open up blue oceans, you must stop focusing on competition. "The only way to beat the competition is to stop trying to beat the competition."[24] Blue ocean strategy is not about taking risks or making decisions without any analytical research and investment. Blue ocean strategy is about reducing risk.

If I took a 50,000 foot, helicopter view of the international industry right now, I would do myself and many organisations a disservice; especially the ones that operate in the office printing and document imaging market. I say this as many multinational organisations have very successfully built their continued survival around technology research and development. Through this focus these organisations have successfully diversified into completely new markets, or those that closely relate to their existing markets, where their intellectual property, capabilities and core competences have allowed them to compete effectively.

From this aspect many firms may be operating in blue ocean markets already. I have to acknowledge that local firms may be unable to enter blue ocean markets in Australia because their parent organisation is testing blue ocean markets in other regions.

For these firms, I will approach the local developments from a broader mainstream market approach by focussing on where the largest amount of strategic activity is occurring and which players are leading this change within the Australian office printing and document imaging sector.

CHAPTER 18:
THE RISE OF MANAGED SERVICES

ARE BLUE AND RED OCEANS NEW INDUSTRY MARKETS?

What are the key industry markets that the office printing and document imaging industry is exploring today? Is this the right move and are they attractive markets to move into? Would you consider them blue oceans or are they just more red oceans? Is this the right strategic move for the players and the industry? Will entry into these markets protect them quickly enough or will the transition be too slow and difficult to avoid the inevitable and ever increasing tidal change of the red oceans where they currently operate?

Let's explore just a few of the new market segments the industry leaders are in, or considering moving into.

BUSINESS PROCESS SERVICES

At present we are seeing a number of the well-known OEM brands either delivering, or in the midst of launching, a business process service (BPS) or business process outsourcing (BPO) service in Australia.

One of the principle aims of a BPS or BPO is to provide the end customer with increased organisational flexibility. Behind this are factors such as reducing production and process costs through eliminating the physical labour required to manage manual and labour-intensive functions like processing medical claim forms. In other words: cost efficiencies.

Other areas of business improvement can include reducing error rates and re-work, improving operational efficiencies and increasing processing speeds and approval turnarounds. In short: eliminate bottlenecks and increase profits while improving the capabilities of that business function or process.

Moreover wider business process services or business process outsourcing can involve the contracting of the operational responsibilities, or specific business process and functions, to an external third party provider.

A number of large organisations such as Australia Post, NEC, IBM, Dell, CSC, Cap Gemini, Accenture and Oracle, to name a few, deliver a variety of BPS and BPO services.

As BPS and BPO can include a wider provision of services, for the sake of this discussion, I will focus on office printing and document imaging. At the same time this approach appears to be a more natural fit for what the industry is best known for. At the same time I would expect there is a level of end client comfort which would suggest that this is a more effective first point entry for players that are new to this industry segment.

From that standpoint, a BPS or BPO provision can be designed to take advantage of the opportunity to transition existing paper-based, back office processes such as mortgage processing, rental bond and tenancy agreements into an integrated digital workflow, data management and processing service.

Business process services has, in some respect, been derived from business process outsourcing, which lost some favour due to the negative connotation that an organisation's labour is effectively outsourced.

However, the business process services strategy is not an uncontested market. There are a number of existing major competitors in this market. I think it's fair to say these OEMs are playing in someone else's backyard. In that respect I would really question whether this is a blue ocean or if these players are just moving into someone else's already established red ocean.

Maybe the view is that it's a very big red ocean and on this point it may be a very valid opportunity. According to IDC "the Australia business process outsourcing (BPO) market (alone) by key horizontals for the 2013-2017 forecast period will grow at a five-year compound annual growth rate (CAGR) of 6.1 per cent reaching AUD$10.7 billion in 2017".[25]

I would like to acknowledge that there is a natural synergy in the sense that the office printing and document imaging industry continues to build its business around reproducing, moving and managing the document or image. As early as in the 1990s Xerox's worldwide slogan was the "Document Company", and whether the document is paper or electronic, the context does not really change. What changes is where, how and what format the user or owner wants the information to be presented in: hard copy or an electronic version.

Business process services is a business model that lends itself to moving to a more paperless world, as ongoing efficiencies and cost minimisation are achieved through the reduction of physical paper-based output and processes. It will be interesting to see how this new BPS adoption process materialises and can be sustained in light of the transitioning of the existing operational and cost model around device hardware sales.

From the industry's perspective the upside of this new business model is that as a BPS or BPO you become more critical to your end customer's business. You start to own the process and in most cases the access to, availability of, or actual data itself. Your value in the relationship grows and it can become very difficult for the end client to jettison you from the account due to the connectedness, intellectual property and knowledge that you have developed as you integrated your business processes into theirs.

I'm not exactly sure who said this first but there is a more common saying today, "that whoever owns the data owns the business". That statement appears to be holding some truth in today's world.

The biggest challenge I see is how today's OEMs convince the end clients that they can do more than manage the transition of paper to digital format as a BPS or BPO service. The end game of OEMs in the BPS or BPO space is to move into more integrated business processes and application management services. It will take time for the OEMs culture to adapt. It will also be interesting to see how the competitive marketplace reacts to such players entering its market space domain. Also at what speed will the end client adapt to this offering from the new players.

DEFINITION AND MEANING OF THE INDUSTRY LEADER SPOTLIGHT

The industry leader spotlight is a guide for the end customer. It is not designed for the customer to take literally as we always recommend a customer perform their own appropriate due diligence.

The industry spotlight is not meant to be an endorsement of an OEM or an industry player either. The Industry leader spotlight is to provide end clients with a view of who is tracking well in a particular area of business, industry market or segment.

Although there is always some brand and vendor supplier self-promotion, some of it is not well supported by execution. I know clients who share this view and it is not contained to our industry alone.

As my disclaimer, customers should always be careful to be balanced in what they read – including many of my own viewpoints. I recommend to each and every client that wants to know more about the areas that I comment on, to do their own independent research. I have learned that there is more than one truth in the world and every situation. As we all know, "it depends".

Unfortunately, there has been a need to qualify the industry leader spotlight as I know some OEMs don't take kindly to an independent consultant and advisor like myself suggesting any view that is not theirs, or doesn't agree with theirs. I fully appreciate and respect their concern.

All I can say is that the mission of this book is to help the end client and the industry to connect and transition more effectively so that both can continue to grow and prosper.

Therefore please take the industry spotlight as a snap shot view only.

INDUSTRY LEADER SPOTLIGHT IN THE BPS

Few players within the traditional office printing and document imaging space currently operate in the BPS or BPO industry segment today. The most experienced player in my view is Fuji Xerox Australia (Xerox).

Xerox has been working hard to transition its business for many years both locally and worldwide. Today Xerox continues, globally, to move to a service led portfolio of businesses. Its services business consists of MPS, Diversified BPO or BPS and vertically-focused information technology outsourcing (ITO). Today services contribute 56 per cent of the total revenue for Xerox based on its 2013 third quarter results.

A significant sign of its commitment to this industry segment came 2009 when Xerox (USA) purchased Affiliated Computer Services for US$6.4bn. This was a game changer.

Fuji Xerox Australia also made a decisive move when it first acquired KAZ Group's business process outsourcing services in August 2007 for an undisclosed sum. KAZ was previously a subsidiary of Telstra, Australia's largest telecommunication company. Their maturity in this space coupled with Fuji Xerox's global services division, which started in Australia around 1996, provides Xerox with a strong competitive advantage over many of its traditional OEM rivals like Canon and Konica Minolta who are only now starting to enter this market.

Canon, in the US, has recently relaunched its managed document services (MDS) program (Canon's version of a MPS program) that includes basic fleet management and advanced document services (which you will hear amount more as MPS continues to evolve past the device level). In addition, Canon Australia launched its own BPS division in late 2013, which may be a version of the US MDS program or direct local play to compete against Fuji Xerox. Time will tell.

HP Australia, Converga and Iron Mountain are all significant players that have operated in this space for some time, continue to develop their BPS models and all focus on or around paper as their starting point. They all provide a number of BPO offerings which, given the increase in and benefits of, cloud-based solutions should up the competitive ante in this space.

For example, locally Fuji Xerox Australia has continued to invest in its capability to help customers design, map and improve business processes across the organisation. Areas of focus can range from accounts payable and creative services through to document services for mail rooms and imaging and document management solutions.

Many of the software solutions (enterprise level, middleware or point solutions) offered to the customer around managing hard copy and electronic documents will continue to grow. From document management systems, content management systems through to server management systems, all are potential parts of the BPS portfolio suite to deliver an end-to-end process for the client.

IT MANAGED SERVICES

A number of OEMs and metropolitan and regionally based organisations have either organically built or acquired IT managed services firms. IT managed services could also include more defined named services such as IT Infrastructure Services (ITIS) or IT Outsourcing (ITO). The strategic intent of such a move can be quite straightforward. However the execution, the skills and capabilities within the IT sector can be very broad.

I speculate that the initial intent for the office printing and document imaging industry in moving into this space was all about moving closer to the decision maker of the IT asset, for example the IT manager or chief information officer. Here the scope of wider IT services opens up and changes the context of discussion while protecting the office printing debate from the more commoditised discussions further down the hierarchical structure.

In doing so, the IT managed services operation would have

increased visibility as well as the remit to take more control and, potentially, ownership of what operated on the IT network. This would protect the MPS services provider against competitive threats in the print and multifunctional devices area.

The second reason that IT managed services became a popular play was the anticipation, around 2011, that a number of traditional IT managed services providers would start to offer MPS as an extension of their already existing contract relationship.

Traditional office printing and document imaging providers started to enter the IT managed services market as both a defensive play and an opportunity to be explored. Although a number of IT managed services firms such as CSC, IBM, Data#3, Fujitsu, Datacom offer MPS, many regard MPS as a poorer cousin. This provided a window of opportunity for office printing and document imaging industry players to enter this space.

The reverse can be seen where it was expected IT managed services firms would enter the MPS space to compete directly against the office printing industry at its own game. However, a somewhat timely technological approach was occurring that reduced the amount of activity or distracted traditional IT firms from entering into the MPS space. This distraction was the ever-increasing marketing presence and excitement of cloud services. Additionally organisations were looking to outsource much of their IT infrastructure. This may have triggered the IT firms to focus on upstream services or specialise around their core offerings.

When I was head of sales and marketing for an Australian publicly listed IT and communications firm in the early part of this millennium, I was clearly told that print was not sexy. It was obvious that blue bloods and traditionalists from within the IT sector would rather go upstream into cloud services than deal with dirty print.

I am sure it is not the only reason why IT managed services firms have not entered the world MPS. However it must be an important factor. My personal experience would suggest the unfamiliar requirement to obtain and manage meter reads and toner consumables replacement, managing billing (including reconciling minimums and excess prints), brand and device configuration, support and experience may have all contributed in some way to IT firms' reluctance to enter the MPS market.

This has set the scene for all the office printing and document imaging industry participants who have established or purchased IT managed services operations to explore the many flavours and specialities that an IT managed services provider can provide. Knowing what you don't want to do is probably more important than what you want to do. The grey or variances can hurt your resources and costs very quickly.

I expect a number of players who have entered the IT managed services market may be coming to grips with this challenge. Their reliance on the existing skill sets may also be problematic in the sense that this is how we have always done things. The growth and development of this new business may not directly correlate with the previous office printing business and therefore the gap of understanding may be a larger concern as the business attempts to transform.

For example, is the OEM, the regional dealer or even MPS firm today an IT managed services business, or are they are an office printing business with IT capability? Which message do they deliver? It may depend on the situation at the time but in my experience, when you have multiple people in the business you need to be clear with your message. It can be difficult to explain and have clarity when you are trying to cover multiple points on a customer touch point.

As they say, the more you focus on it the more it expands.

INDUSTRY LEADER SPOTLIGHT IN IT MANAGED SERVICES

A number of major OEM players have moved into IT managed services. To date I have been most impressed with the acquisition Lanier Australia has made. Lanier Australia is a subsidiary of Ricoh Singapore (Japan) and both Ricoh and Lanier have competed against each other in the Australian market up until March 31st 2014. Lanier Australia will now be integrated in Ricoh Australia, which effectively calls time-out on Lanier Australia.

Lanier Australia purchased Inspire IT in early 2013 and has put this acquisition to good use. Prior to the Lanier acquisition, in October 2012, Ricoh Australia invested in IMC Communications to enhance its Managed Document Services capability, so both firms have made strong moves into IT services. The new combined entity will certainly make a statement as it moves into this space.

Certainly a larger group of industry players that should be noted in this space including HP, which would be the strongest global player in the market from an IT services capability. Other firms such as Fuji Xerox Australia, Dell and some regional players like Datacom, Data #3 and Viatek (which recently acquired CN Group) are all currently positioned well. CSG may be well positioned to re-enter the space if it chooses to, having sold its technology solutions business to NEC Japan in early 2012 for $227.5m.

Some firms, such as Canon in the US, are setting up go-to-market plays like Print IT Infrastructure Management Services (a move to capture the Print Server Management Service). At this stage I have not seen this carry over to the Australian operations.

If I was to crystal ball the next minimum level of core competency or business capability the industry would need to have available when going to market, it would be around IT infrastructure and IT managed services.

The offering would be based on the broader aspects of a customer's IT Infrastructure requirements, leveraging up from print and imaging. The reasons for this view are varied and include:

1. The interrelationships, technical integration and intersection of the industry's existing solution set is the most suited and aligned of any existing industry outside of the current IT managed services provider.

   I say this as there continues to be an ongoing fragmentation of the IT managed services industry in Australia in general, with a limited number of large national IT managed services organisations operating.

   With the advent of more innovative, attractive and exciting technology and platforms to deploy, traditional or purist IT service providers are somewhat less focused on what they consider basic and less sexy – including print, imaging or the processes and applications around this arena.

   More to the point, a number of aspects of IT infrastructure now look less exciting, more competitive and in some respect less important and less valuable to a number of IT managed services firms.

   This does not mean that each OEM or office print and document imaging player that goes to market requires all the deep technical skills in-house. More important is that they take ownership and responsibility for managing the requirements and outcomes required – very similar to what they proposing around MPS.

2. The second point is that end clients continue to shift from capital intensive IT purchasing to a more agile, utility-based or "as-a-service" model. This, in itself, is fundamentally reshaping and changing the current IT services model. In doing so it impacts and changes the traditional or conventional manner of managing

IT vendors.

This means that the timing and relevance of new significant industry or business partners that can couple the business, technical and investment requirements at a business and technology infrastructure level will be the perfect opportunist play for an industry player and customers who are increasing their appetite for continued strategic vendor partnering.

Enter the office printing and document imaging industry...

SOFTWARE SOLUTION PROVIDER

Virtually all industry players can, and will, offer print software solutions today. Some will offer enterprise level software, either theirs or someone else's. However, most software offered today still encompasses, or is in close proximity to, either print or imaging capabilities. From this consideration all must be seen as software solution providers.

Moreover, many specialist software players have entered the office printing and document imaging industry as support technology to the device (monitoring tools) or printing process (document conversion and automation tools) or to add value through extending the capabilities of MPS such as big data analytics.

In this fashion we currently see virtually all brands developing hardware and software around the business requirement for mobility printing and data security. Increases in tools that provide increased data analytics will be the next big wave that helps the industry build and develop increased customer intimacy.

What we see today is a software lifecycle that mirrors and

supports the office printing and document imaging ecosystem. One specific example is the various offerings of device management software to the market. The software is either native to an OEM vendor or an alternative such as Print Audit, FM Audit and Papersoft, the most prominent three in Australia. Depending on the software installed, device management may include features like as device alerts, supplies alerts and usage tracking.

Software that enhances an MPS offering for customers can come from an OEM. More commonly we see an increasing number of software solution providers that position themselves as independent as demonstrated by their ability to integrate with multiple brand platforms.

This is a strong consideration point for end customers that fear being locked into a single brand's technology platform and, hence, their devices. Customers generally prefer open architecture software that works with their choice of product or solution vendor of choice.

As an independent industry advisor to CIOs and CFOs, I continue to hear a consistent message from clients that they want the freedom and flexibility to choose a brand, product (hardware or software) depending on the business requirement at that time, a solution that is fit-for-purpose. In this respect they give far greater importance to products and software solutions that can be flexibly and seamlessly integrated into their operational and infrastructure environment.

For example, pull or push printing (depending on which service provider you are talking too). Pull or push printing (see Figure 2) is a term used to describe a print function that enables a user's print job (file) to be held on a server until the user performs some type of authentication at the actual multifunctional device or printer location which then releases the print job to the device.

Touted as a labour-saving, resource-conserving, security-focused and cost-saving feature, pull or push printing describes the capability for users to print to a shared device anywhere in a building, or a campus regardless of whether you are visiting from another site or work locally. You retrieve the print job by entering a PIN or, even more common today, waving your security card at the selected device.

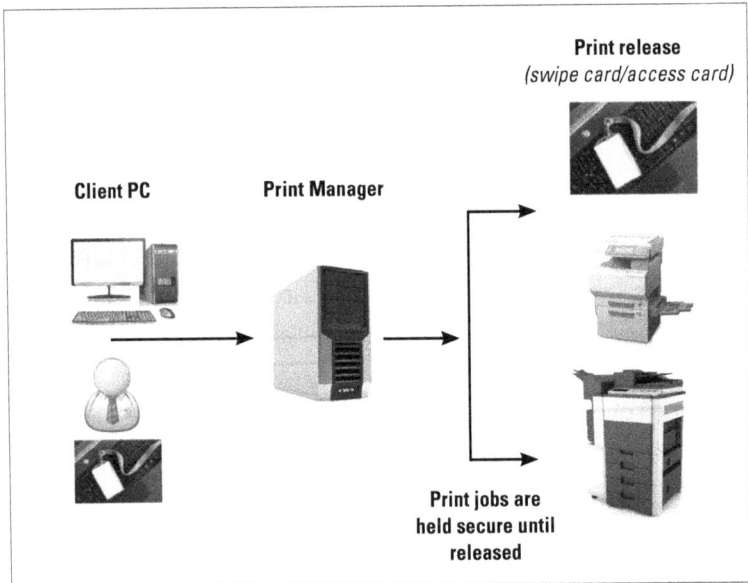

Figure 2: Pull or push printing

As indicated pull or push printing provides users the flexibility to print from any device they choose. In a consulting practice, for example, a user may live and operate out of the Melbourne office in Australia. When work takes them to the Sydney office, having swipe card access to the office, the software and the device they choose is both vital and efficient to the user and consulting practice.

A word of warning, though: in some cases we see areas of incompatibility when deploying such solutions. This can occur in relation to the access or proximity card being used at different locations or departments, the way the software was deployed or incompatibility between the device hardware, the proximity card, card reader and the software.

In this respect, software technology that is limited, restrictive to brands, platforms and device models, is unfavourable and is unlikely to be recommended by the decision makers. In fact I have been told by many customers, on many occasions, that they always prefer software that is more open and flexible as business is increasingly focused on agility. Large, cumbersome and closed architecture is not their preferred option.

On this point, there is a plethora of software to choose from an assortment of players that bonds today's office printing and document imaging space together. This space will continue to grow as businesses continually strive to extract value from hardcopy and electronic documents.

This will bring into focus software tools around document conversion, automated workflow and data analytics that will drive the efficiencies of the users and the business. Data analytics will provide businesses a new level of insight into their customers and with this data, knowledge and insight it is expected that businesses can become even more intuitive of customer requirements.

This brings us to the point of nexus that exists today between both the physical and digital worlds. The scanner, the on-ramp to the digital world – in fact the location that makes that multifunctional device (or the single-function scanner) an even more intelligent information hub for business users – is where the start of many future digital business processes will commence.

Utilising scanned document images after they have been scanned through a multifunctional device or printer or dedicated scanner is now the core focus of many customers. In digitising the image from the physical paper document, organisations are now interested in storing, searching, sharing, securing, analysing, moving and extracting the data value and information content to support increasing functions and requirements of the business. This is becoming a focus as the maturity of organisations to better utilise technology platforms, improve business process efficiency, integrate software applications and utilise business intelligence and big data are now major reasons why business wants to move information from the physical page to an electronic format. Yes this all stems from that information hub in the corridor that we call a multifunctional device, printer or scanner.

The OEMs today are certainly pushing hard for the integration of the device to the cloud. I expect we will see many different offerings and programs of this type of integrated technology from OEMs as well as other existing and new entrant players.

INDUSTRY LEADER SPOTLIGHT IN SOFTWARE

Although there are industry leaders who continue to build the ecosystem around their product lines and business services, such as Canon, HP, Fuji Xerox and many others, one has focused more intensively on software. That brand is Lexmark.

Globally, Lexmark appears to be making a bolder transition into software with over eight acquisitions since 2010 including Perceptive Software, a process and content management solution; PACSGEAR, a leading provider of connectivity solutions which integrates medical images with existing picture archiving and communications systems and electronic medical records systems; Twistage, a cloud software platform for managing content; and Saperion, a German based software provider of enterprise content management software.

Closer to home, Lexmark purchased Australia's ISYS in 2012, an enterprise search software that allows organisations to deliver federated search, text mining, and mobile and embedded search capabilities across a large range of formats, languages and platforms.

Although I have spotlighted Lexmark due to their level of public acquisition activity, it is important to note that many international firms such as Canon have large investments in continuous research and development programs (Canon Inc., for the 28th consecutive year, has ranked among the top five U.S. patent holders). Therefore firms such as these continue to fly under the public radar of progress.

There is no doubt that some industry peers and observers may think their organisation should have received more of a worthy mention; my apologies for their omission. As I have previously stated, there are many brands and service providers across the spectrum of office printing and document imaging that deserve mention, let alone organisations that are operating on the fringe or making plays into the sector.

I have deliberately not discussed, in any real detail, organisations from neighbouring sectors that cross over to our industry such as the tier 1 IT services firms such as IBM, CSC or Fujitsu, all of which operate in our industry. It would be an endless task to include surrounding industries for special mention. My focus for this chapter has been more around the epicentre of our industry.

As one last note on this topic and chapter, I appreciate that I have not given a deserved mention to many of the innovative and successful businesses in regional Australia and New Zealand that continue to invest in and around this technology space.

In many ways these players are not doing this because they choose to be innovative, but more because their customers are asking them for help, guidance and to protect their business in their local regions. This relationship of trust and support goes further, and is far more important, in small business communities than many people realise.

Many of these smaller players in Australia and New Zealand

require a chameleon approach in assisting and working with their customers. They have to punch above their weight and they don't have the luxury of, and access to, resources like the big end of town has. My personal acknowledgment to your continued work.

In summing up, the premise of this chapter was about the challenge facing the industry, where it operated today and where it may have to operate tomorrow for both big and small players. The mention of any organisation (brand) was purely as a reference point to illustrate where a focus of transition is moving.

Moving on, what does continue to shine through, and linking back to our framework of blue and red ocean strategies, is that each player operating in the office printing and document imaging industry today is trying to build a legacy for tomorrow.

The ability to move into newer markets and industry segments appears to be more common from an acquisition play rather than through organic growth. The speed to market, domain expertise required and the focus on resources are some of the attributes that make it more efficient to buy the talent and know-how. This is especially true when you need to continue to focus on your core business without the distraction of growing and changing a potentially larger part of your business.

We currently see the office printing and document imaging industry reinventing itself in a number of ways. Each organisation is trying to shape a new opportunity that will protect it from an ever-growing red ocean. Most firms are investing in potential blue ocean markets through acquisition, however this may change as the level of maturity changes within their newer industry model.

Looking from the outside in, it would appear that these so-called blue oceans may actually be someone else's red oceans. Therefore, as this chapter started out to really question how the industry is adapting, is it doing enough to develop the next blue ocean or uncontested market?

I would suggest that some players are very early in their market or business maturity cycle. Some are certainly trying to do one, a few or all of the industry segment plays that I have mentioned as well as others that I have not mentioned. I expect the industry will continue to change and this continues to demonstrate the strength of this industry as it transitions away from its office printing core business to a wider all-encompassing managed services offering.

CHAPTER 19:
SERVICES, SERVICES AND MORE SERVICES

THE INTERSECTION OF THE ACRONYMS BPM, BPO, BPS AND ADS

I have included this topic as a deeper drill down on a market and industry segment that is currently very fluid and continues to overlap its scope of business practices and business services. Due to its relevance I have decided to provide a high level snapshot of the intersection of services and terminologies used.

Explaining the cross-over point of frequently used terms and acronyms such as business process management (BPM), business process services (BPS), business process outsourcing (BPO) and now advanced document services (ADS), just to name a few of the buzz words, can be very confusing and unclear to many.

As I say this, I see an increasing intersection of such terms, which are not necessarily new to some of you. In fact, some may be dressed up differently, under a new disguise, as their scope of meaning, capability and desired outcome has shifted.

Based on what I am continuing to see and hear publicly, as well as what I am being told (off the record) I thought it was important to highlight where these terms come into play and where their focus is, particularly from an Australian perspective.

The following is not designed to be detailed to the point of boredom. It is really a heads up to the language, terms and what these terms can mean below the surface. It is intended to capture how the industry is broadening its capability, where these capabilities have materialised from and where they are shifting too as the industry continues to broaden its scope.

In relation to our core industry, we see BPM experts in Australia focussed on identifying, mapping and modelling how current manual paper or labour process can be made more digital friendly, operationally efficient and cost effective. In fact, reducing the cost of processing is probably a more accurate assessment of the client's objective.

This can be achieved through a range of activities such as digital form creation, digitalising paper-based documents, electronic workflow improvement, reducing manual labour processing (such as time and motion studies of processing a form or document) and automating approval processes that integrate and leverage other business systems and applications.

More simply, and in most cases, they look at reworking large and labour intensive paper-based processes in corporate and government business. This could include such work as property rental and bond applications, credit checks and approvals, home loan mortgage and insurance documents as well as credit card and banking approvals.

In most cases, their major aim is to remove human labour, which is the most expensive part of the back office function for many businesses. The other major benefit is that it makes the business more efficient through faster processing. This delivers obvious additional benefits to both the business and the customer such as increased transactional activity that improves revenue flow and improved customer satisfaction and patronage.

BPM can be the start or part of a larger BPO strategy or operational contract. BPO is really taking ownership of, and responsibility for, a defined end to end process or an entire business operation at a cost that allows the end customer to focus on their core business activities while assigning the expected and measurable outcomes to the outsourcer to deliver. In some cases, the end client may feel that this process is best delivered by an external third party. In these situations the outsourcer may improve the previous business process or operation to a point where the end client sees value in bringing it back under their operational and business control. The BPO may reengineer the process or business function to the point that this reengineered operation is now perceived to be best of class and adding value to the businesses value chain.

On the BPS aspect Fuji Xerox Australia has been running a business processing service for well over five years and Canon Australia officially launched its BPS version in late 2013. These businesses are heavily aimed at either owning, operating and or managing the entire or part of the process (process, technology and labour) depending on the scope, experience or risk of the service required. They integrate with customers either through an insourced, hybrid or totally outsourced provision services model.

Fuji Xerox Australia has certainly leveraged its Global Services Group, which was formed in around 1997, and has been building a client base around imaging, document management, mail and scanning services, production print and more since then.

With this said, many end customers would still consider the office print and document imaging industry to be relatively new to this market. Although new they certainly are well positioned and have the financial commitment and technical capability to play a more defining role in this somewhat related industry and market.

I expect that OEM brands such as Konica Minolta Australia,

Canon Australia and Ricoh Australia will make a good play into this growing market as the opportunity for a services relationship continues to gain momentum.

Advanced document services is a new term, recently crafted by Photizo Group which, in my mind, is really designed to focus the office printing and document industry away from the transactional device play and make it part of a much larger MPS strategy. It's preparing the industry to ready itself to migrate away from its current paper-based income stream to a new income stream.

ADS is about helping clients transition from paper-based documents to a digital-friendly format. Document automation, combined with document workflow processes, can be part of a larger business opportunity to integrate their extended services offerings into a valued and trusted partner that is no longer considered necessarily transactional in its relationship.

ADS will encompass many of the aspects that were described in Chapter 18 regarding the rise of managed services. ADS is about helping transform a customer's business while creating and extracting value for the client. ADS is about moving to the next level in the client–value proposition. ADS centralises its view at the automation, workflow, process improvement of data and information within the business and then leverages data analytics to help to drive more informed decision making and insight.

Irrespective of all these new terms, business services and market hype there is one thing we have to remember: we will not be transitioning all paper-based output to a digital format tomorrow or the next day. There will still be a requirement for printing in many offices large and small. Printers, MFDs, and MFPs will properly reduce in unit numbers over time and maybe more significantly than we think. Time will tell on that score. However, the focus will now be on the most repetitive, costly, time consuming and labour-intensive areas of document handling. This is the area under the most threat to transition from the current printed page to a more streamlined digital process.

The key fundamental driver behind this transition is that business has to reduce its bottlenecks if it generally wants to improve its overall performance. Taking paper out of a process in most cases makes the business more efficient and, at the same time, reduces its cost.

Business wants to speed up the rate that transactions come into the business and reduce the costs of each transaction. The printed page needs to seamlessly fit within this framework.

In summing up, we are starting to widely acknowledge that there is a decreasing requirement for holding paper itself. It's now recognised that it is more important to hold, and utilise, the content on the page rather than the paper itself. The "content" is valuable, not its mode of transport – just like the horse and buggy, the mode of transport is changing.

Paper is a limiting factor when there is an increasing requirement and demand for access in a world that expects mobility, portability and availability 24x7.

MORE SERVICES TO COME

Although these last two chapters have made specific mention of a number of key areas of focus where the office printing and document imagining industry are clearly building new business models, skills and capabilities within a services-led delivery framework, there are other areas or business offerings that I have chosen not to describe in more detail.

I think it is valuable to at least cover a few technological and business service offerings.

Cloud printing services

Today I see two types of cloud printing or cloud service options being discussed by both the customer and the industry.

The first is currently available via mobility printing which is, effectively, one form of a cloud printing service.

The second is a much larger debate and I know a number of CIOs have put their case forward for a true utility printing service model. Simply stated, the user pays only for the service when it is used or required. In fact, one of the CIO round tables that I conducted in Australia in 2011 for the Asia Pacific MPS conference on behalf of The Photizo Group raised this specific point. Phil Hurley, CIO of Australia's third largest private hospital group, raised this issue in front of a packed audience that were very much industry players: OEMs, IT services firms, resellers, dealers, agents, Hybrid players and so on.

The debate was robust. Mr Hurley delivered this comment to the packed audience: "I do not want to buy or own the devices and I do not want to sign up to a three or four year rental or page per print contract either". Many of the industry players were quick to voice the same statement: but who will pay for the devices? Mr Hurley's quick reply was, "that is your problem". Now, this was three years ago and the noise has just got louder since then.

Before I continue, let me explain in a little more detail what a utility services model could look like. From the customer's perspective a utility printing services model is the ability to switch on or off the service provision at any time with no real adverse penalty to the end customer. A service of this type allows the customer to make more agile decisions about their business operations or business requirements at the time. In fact it's very similar to switching on a power point or light switch at home. You pay for what you use; if you don't use it there is no cost.

The major stumbling block with this concept to date, is that service providers have to fund the capital amount to provide a total print solution. Today the cost of funding such hardware and software assets makes the cloud printing service uncompetitive or unattractive to the marketplace.

Additionally, if the service provider was to make this cloud printing service offering attractive to the market, the service provider would, in almost all cases, have to fund these assets over a longer set contracted period to reduce the cloud printing service fee. Effectively, this would tie the customer to a longer term contract, effectively moving the customer away from a true utility printing services model and back to the current page per print model that is currently offered.

Scanning services or, in the new world, Scan-aaS

Scanning services has been around for a good decade now. However it has always been geared around large scanning volumes. In fact many players who offer some form of business processing services would include scanning as part of its delivery capability just through the necessity of moving paper-based documents into a digital format.

However, we will see scanning services come down to smaller businesses as scanning solutions, cloud offerings, network bandwidth and areas such as customer and business intelligence play a more important part in beating the competition. The main drivers for the shift toward smaller business is the reducing costs of technology and the requirement to unlock valuable physical information from the page.

"as a service" (aaS)

These days everything can be retitled or delivered as a service. I expect some of the industry players to make use of these terms in the coming years. As they build their services maturity and establish experience in delivering services they will pick and deliver the appropriate services aligned to their customer.

In the future you could be seeing the office printing and document imagining industry marketing and delivering such services as infrastructure as a service (IaaS), software as a service (SaaS) and application as a service (AaaS) just to name a few

In summarising this chapter we wanted to address and highlight areas of the market where the OEMs of the office printing and document imaging industry will become more active as they continue to move from a product-led transaction business to a service-led business capability.

This newer area of focus will remain challenging for both existing and new players as they try to establish and maintain their credentials around capability and capacity. Challenges such as security or data sovereignty, back up, redundancy and disaster recovery as well as delineating the scope of operational infrastructure and their services (such as cloud services) will become important requirements for both end customer and as the service provider.

Section 6:

WHY THE CUSTOMER NEEDS TO AUDIT

CHAPTER 20:
LOOKING UNDER THE HOOD WITH A COLLECTION OF CASES STUDIES

In the following chapter I thought it would be worthwhile to share with end clients the value of three different case studies and their findings. The case studios are from different industries: finance, health and a semi-government institution. Each case study focuses on three different areas or business drivers based on the end customer's scope: MPS benchmarking, determining the size of the prize and increasing the transparency to the executive.

The point of these three case studies is demonstrated not only in the value of the exercise both in financial terms but also in terms of good ongoing governance.

The three case studies are high-level overviews only. They were written to provide all executives and managers a quick insight into how gathering the right data can be your best friend when making informed decisions for the business. It was also written to provide a spot light on how printing costs and contracts can differ and change and, when this occurs, what impact it can have on the business over time.

CASE STUDY 1:
DETERMINING THE SIZE OF THE PRIZE

The client is one of Australia's largest privately owned health organisations with a turnover that exceeds AUD$250 million. It manages an estimated 1500 staff over approximately 14 centres, which include a range of hospital and aged care facilities across the Eastern states of Australia.

SITUATIONAL BACKGROUND

Our client was the company's chief information officer who was relatively new to the organisation, had extensive senior executive management experience and IT operational experience. The organisation was a private equity play with a simple strategy of growing the asset value of the business.

From an IT perspective any strategy deployed had to increase the asset value of the group. In our experience, many senior executives who sit outside IT see IT as either a cost centre or a cost of doing business and not well positioned to contribute to the asset value of the business.

The task for the CIO was made even more challenging by a small IT budget (less than three per cent of turnover). Technology had been purchased, deployed and managed in a decentralised manner and a cultural shift was required to move their ad hoc IT operations into an enterprise-grade capability. Looking back, this was the perfect client with the perfect mantra: "Do more with less".

Fortunately the client was astute and I'm not saying that because they chose us. They had a clear vision of what the big picture looked like and the process to achieve it. They were transparent but not naive and they had a true compass towards what they had to achieve.

Once we were given the approval the CIO shared more of the lower tactical challenges and where our role was required. The IT operation was decentralised in its technology deployment because it had previously catered to several different owners. Therefore, the first task was to centralise control. As part of this first step, we were tasked to provide transparency to the CIO in relation to print and multifunctional devices across the network. Stage one was hardware-centric, centralised cost control and preparation for stage two.

OUR ROLE: STAGE ONE

The client was very clear on our objective: "What is the size of the prize?" They wanted to know and did not want to us over-engineer the objective. Was there a quick and measurable win?

We agreed a plan with the client to segregate a group of devices that were under several different contracts. Although we took in a bigger picture audit we were selective with the audit scope and only audited networked devices. The company's extensive collection of USB (non-networked) devices remained outside the scope.

The audit of networked devices accounted for 288 devices. The CIO was confident from initial fact-finding there were a similar number of USB devices. That equated to just under 600 devices. The first realisation was that, with a staff of approximately 1400, there were approximately two people using each device – even given the geographical challenges of multiple sites. Device deployment and utilisation rates are generally higher in the health industry so this client was still way off trend.

ACTIONS

Over an eight week period we monitored and reviewed all networked devices and substantiated print volumes. We also validated devices against IP address and married up devices against
finance to identify whether each device was on an individual or
master contract. We then audited invoices against minimum print
volumes, fixed or variable print charges, print increases and excess
prints over minimum volume thresholds and invoice billings.

HIGH LEVEL FINDINGS

We established the following findings, which unfortunately are
not uncommon:

- Of the 288 networked devices audited we were
  able to establish that they were made up of 12
  different brands as illustrated in Figure 3: Vendor
  range and device totals
- The 175 devices that were part of Phase 1 were on
  seven different contracts
- These 175 devices averaged 625,000 prints per
  month and just over 7.5 million prints per annum
- Total print volumes for networked devices was a
  75:25 ratio split:
  - 75 per cent of the total print volume was
    produced in black and white
  - 25 per cent of the total print volume was
    produced in colour.
- Represented as a cost, the ratio was the exact
  opposite:
  - Black and white prints represented 25 per cent
    of the costs of printing

- Colour prints represented 75 per cent of the costs of printing.

VENDOR RANGE & DEVICE TOTALS

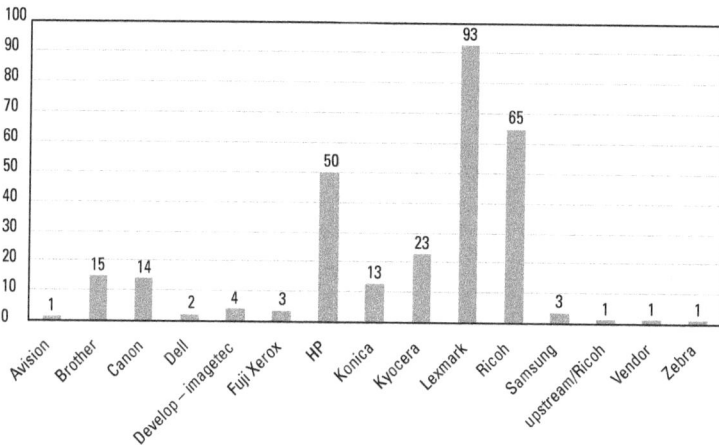

Figure 3: Vendor range and device totals
(Note: Avision and Zebra were not included in vendor totals. Upstream/Ricoh was included in Ricoh vendor total and Vendor was included in Fuji Xerox total)

OUTCOME

After a detailed analysis we highlighted a number of disparate and contradicting areas of their existing contracts. We provided financial modelling which calculated current costs based on average print volumes in both black and white as well as colour.

Although the client recognised the larger opportunity of savings they did not want to boil the ocean and remained focused on the objective that we agreed

Therefore, based on the scope of works for stage one and

the objective for the client to actualise the size of the prize, we established that they could save more than $450,000 over the next 48 months.

The CIO commissioned us to commence savings realisation for stage one and commissioned our business to review stage two in the same manner as stage one. This accounted for another 313,000 prints per month.

The client requested that we take on the role of independent trusted advisor to assist in the process of future contract management. As we do not earn income, endorsement or commissions from any brand, device or service contract we were uniquely positioned to support the customer's request.

The benefit to the customer of engaging us is that we are not consumed by the device or unit numbers or print volumes, which typically drive print industry behaviour. Our ability to work with both end customers and the clients' recommended suppliers allows us to achieve the savings required by the client.

We managed this pricing framework transparently with both the end client and supplier to achieve the best outcome for the client. Both parties realise the importance of maintaining a fair and equitable balance and a level of commercial viability.

Far too many times we have seen suppliers cut pricing too aggressively and it ultimately impacts the client's business later in the contract term when the transition cost of switching can be prohibitive.

FINANCIAL SAVINGS REALISED

Stages one and two realised total cost savings well in excess of $500,000 over the next 48 months.

CASE STUDY 2: *PROVIDING INCREASED TRANSPARENCY*

The client was a global not-for-profit organisation operating in every state and territory of Australia. Our particular end client was the company's CIO who was responsible for 200 physical locations in metropolitan and regional New South Wales and Australian Capital Territory.

SITUATIONAL BACKGROUND

The company was in the early stages of a large IT transformation program. As such they were required to provide a roadmap for their future office printing requirements.

This gave the executives across business, IT and governance the opportunity to confront a number of increasing concerns which included a lack of device visibility across their locations, device investment and contract approval policy across these locations, existing contract consistency with existing vendors, standardisation of device technology and platforms and an accelerating cost structure with regard to their office printing.

As part of the project establishment we worked closely with the CIO and manager, business and governance to ensure that the process and business review could be presented to the board committee for approval.

PROJECT PHASE ONE

Our objective for phase one was to establish a current cost base and a standardised platform for technology deployment. This needed to include both user and business requirements that provided for capability and capacity for each location or category segment prior to going to market. Additionally the business wanted a review and audit existing contracts and market benchmarking prior to exploring the market.

ACTIONS

An extensive audit of existing contracts, invoicing and billing illustrated a number of major inconsistencies with both capital configuration pricing, financed rentals and service and maintenance contracts.

It was agreed that, as part of their existing desktop roll out, they would conduct a physical audit to either eliminate or record USB (non-networked) devices while we were tasked to audit all networked devices.

In parallel to these activities, a user and business requirements survey ascertained key business and user requirements for office imaging such as printing, scanning, faxing, and copying.

HIGH LEVEL FINDINGS

- Our audit of networked devices indicated that the 323 devices were made up of 15 brands.
- Astonishingly, of the 323 colour and black and white devices there were 125 different models of printer, MFD and MFPs (see Figure 4: device vendors and unit numbers for colour and mono printers).
- 170 devices, or 53 per cent of the fleet, were colour-enabled and the remaining 153 devices, or 47 per cent, were black and white.

- The top three brands represented 66 per cent of the fleet (two thirds).
- Approximately 62 per cent of the devices were A3 capable.
- Colour prints represented 30 per cent of print volumes and black and white the remaining 70 per cent.
- Colour represented 75 per cent of the total costs and black and white the remaining 25 per cent – almost a reversal of the print volume numbers.
- Total prints for the 323 networked devices were 1,241,881 per month (just under 15 million prints per annum).
- The ratio of printers to desktops equated to approximately 8:1 (one device for every eight people).

DEVICE (Vendor/Unit Numbers) FOR COLOUR & MONO

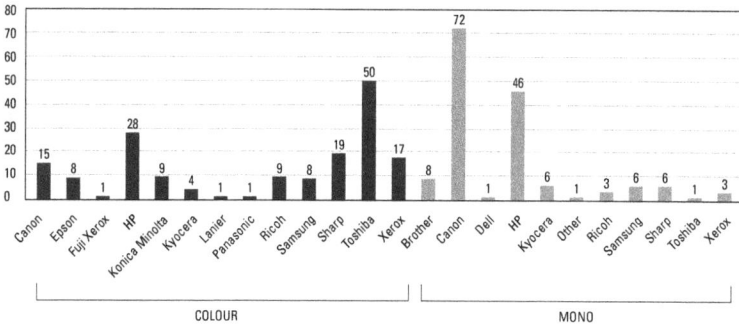

Figure 4: device vendors and unit numbers for colour and mono printers

OUTCOME

Our contract audit, service and maintenance review, reconfiguration of consumables and warranties, model rationalisation and standardisation and new refresh program for both metropolitan and regional location produced a cost savings of more than $42,000 per month.

This did not include cost savings from decommissioning the USB devices and strategies such as automatic double siding to reduce use of paper and consumables.

CASE STUDY 3: MANAGED PRINT SERVICES BENCHMARKING

The third case study was an MPS benchmarking study for one of Australia's largest and most recognised financial services organisations. In fact it was the first independent MPS benchmarking analysis and report ever to be conducted and produced in Australia.

The customer has more than 5000 staff worldwide, around AUD$179 billion under management and operates a number of brands across three distinct business units. The business was preparing to make a large acquisition and wanted to review its current partnership program across a number of businesses and IT functions.

SITUATIONAL BACKGROUND

Our end client was the senior commercial manager of the finance organisation's Alliances Group. In addition there were a number of key stakeholders that included a commercial manager (finance) and a benchmark practice manager from a third party IT integrator and solutions partner.

The business had made significant changes over time to its existing fleet management program as well as its outsourcing partners and carried well over 250 networked multifunctional and single-function devices across Australia. Although happy with progress, relationship, services and commercial arrangements with the existing partners, the business still required an independent and audited analysis due to the businesses awareness of market fluctuations.

PROJECT PHASE

The project ran over four weeks and included market and competitive analysis, financial modelling and industry benchmarking against relevant or related market industry segments. The benchmarking report would also provide a capability assessment of the services provided within the scope of the existing managed services contract. This assessment would also detail costing against services provided and areas where cost savings and paper-based business processes could be improved.

ACTIONS

Working with all key stakeholders we validated and synthesized the data sets from a range of sources. We extracted data from the current partner contract and partitioned data that we concluded was outside the contract to ensure the services were comparable on a like for like basis.

Working globally in the first instance we developed a best practice methodology for ensuring measurements were consistent, meaningful and related to the Australian marketplace. We then worked locally by selecting and interviewing key segment customers to obtain an insight to their current metric pricing contracts.

The process included inputting and calculating pricing metrics to formulate both an average as well as high and low level ranges. Data analysis and comparative benchmarking did not include customer contracts or data analysis outside of 18 months (service pricing older than 18 months was considered out-of-date for this benchmarking activity). In short, we focused on customers from the same, or similar, vertical industry sectors and organisations with similar dynamics or elements such as population of desktop users, number and diversity of locations, revenue turnover and current printing devices deployed.

HIGH LEVEL FINDINGS

- The range and quality of services that the customer received for their MPS (in-scope) were calculated to be in the top five per cent of the Australian Market of similar customers with similar fleets across multiple locations in Australia.
- Cost benchmarking of the customer's competitive blended rate for multifunctional and single functional devices, both black and white and colour, were in the top quartile for pricing competitiveness.
- Cost efficiencies were in the top quartile due to the existing MPS partner managing the organisation's help desk service.
- Overall services uptime, service management and user monitoring were in the top quartile for both cost efficiency and customer or user satisfaction.
- Contract complexity for the customer was considered to be low and consistent to manage.

OUTCOME

The client was comfortable that its current MPS partner has continued to demonstrate competitive market pricing and therefore there was no reason to change or go out to market prior to the new business acquisition. More importantly, the customer saw additional value in combining helpdesk services with its MPS contract to capitalise on its easy management and efficiency. It also provided the customer less overall IT management responsibility and therefore proved to be better value for money.

Disclaimer: Due to the confidentiality and nature of this customer and the competitive analysis of this benchmarking report, no financial or specific details can be provided.

CHAPTER 21:
FIVE KEY FACTORS TO DEMYSTIFY A PAGE PER PRINT CONTRACT

When organisations look to reduce or consolidate their cost of doing business, the first steps are usually to itemise and review areas where their current business costs are either unknown, uncertain or starting to increase.

In many instances there are proven business best practices or quick win opportunities to achieve cost reductions or consolidation that many companies can manage on their own. However, in some cases, businesses may require a deeper industry understanding or proven processes to provide increased transparency and analysis of these costs. Sometimes it's difficult to know how suppliers break down costs and charge them to the business. In many cases, existing contract pricing and invoicing charging conditions change and businesses can be uncertain why and how increases have occurred.

Many customers we meet either need to go to market for a refresh of their device fleet or are looking to implement a MPS program that will include contracted elements with one or more suppliers. In other cases, some customers are trying to make sense of an existing contract that they have inherited and their concerns are raised due to inconsistencies in the contract logic of existing suppliers.

In all cases the same problem arises: that the page per print (PPP) contract is not working for the client or they feel that the supplier or suppliers are not as transparent as they should be, raising concerns about trust. Some clients simply feel that they are not getting the best deal but find it difficult to compare and benchmark the contracts as the key elements are difficult to break down.

To assist clients in senior executive roles across finance, IT and procurement I have prepared five key factors to demystify a page per print contract.

INTRODUCTION

In Australia we have seen many technology contracts that allow the customer to invest in, use or have their devices maintained as part of a contract provision. In the office printing and document imaging industry many different suppliers, be they direct brand manufacturers, resellers and dealer channels, offer a number of different programs, all differentiated from each other with different names.

Without listing all the different hybrid product and program names, we have simply called this the page per print (PPP) contract. However, some would know it by its old, analogue name, the cost per copy (CPC) contract or a fusion of both, the cost per print (CPP).

THE FIVE KEY FACTORS THAT YOU NEED TO KNOW

1. Be clear that this is a financed product

Many clients that we meet feel that they have been misled by the vendor or supplier to think that this is more of a utility model (pay as you go) and that they are only charged usage by paying for the prints they produce each month. The devices that they use to produce the prints are only a function of the volume they produce, therefore the asset importance (ownership) within the contract transaction is downplayed. The customers are more of the view that the asset (or device) is not as tangibly important because they are not buying the device – they are buying the print volume associated with the contract. In their minds they can return or replace devices at any time without a penalty of any kind.

A finance product, such as a page per print contract, comes with a commitment to the physical asset used. Any variations to the original contract may come at a cost to the end customer. Although the appearance of a variation may not have a material cost difference in the initial stages, it's towards the back end of the contract when the contract thorns starts to show themselves. This causes much grief and angst for the customer.

The relevance of a finance product must not be understated. The fact is that virtually none of the manufacturers, resellers or dealers provide finance directly. Despite what they say, what they mean is that they will provide finance by outsourcing finance to their finance partner, which may be a different company even if it has a similar name.

Their finance business is, in most cases, a separate entity outside the sales, service and marketing operation. This applies to global brand manufacturers as well as channel and dealer resellers. On the surface there is nothing untoward with this arrangement. It provides a level of protection for the finance entity as well as the sales, service and marketing operation. I will share one of the scenarios I have seen so that readers can better understand the potential complexity of this process.

If an end customer on an existing page per print contract decides to stop paying their contract due to dissatisfaction with the level of service or frequency of device breakdowns, the risk to the finance company is far greater than the risk to the sales organisation. The quick answer is that finance collects the income and pays the service portion back to the sales company. The debt exposure is between the customer and the finance company so withholding payment can sour the relationship with the financier and not actually achieve anything with the sales company.

I expect many sales and finance representative would say this should never happen, but it does. Frequently. Unfortunately this situation is most visible when it becomes evident that the relationship between the parties has deteriorated past a point of distrust and frustration.

Some customers honestly believe that this disconnect between the agencies involved is a strategic counter measure to customer recourse over issues with any component of the PPP contract.

In summarising we have worked with a number of customers who have been constrained in their pursuit of improvements in the contract status and obtaining their desired contractual outcome. In many cases we help them to identify the appropriate point of communication for achieving effective mediation as well as improving the relationship with the vendor or even helping the customer negotiate with the vendor directly to achieve a more favourable outcome.

In all situations the number one failing is a communications breakdown. This is due to the customer having to navigate with the salesperson, the sales management teams, various accounting and finance teams, the service teams and any third parties that are involved.

2. The three ingredients of a PPP contract

The PPP contract is a finance contract and operates with three key ingredients:

i. **A contracted term (for example 36, 48 or 60 months)**

A page per print contract, like any other finance contract, requires a starting and ending date. Office printing contracts are generally financed for 48 months though, in the late 1990s, 60 months was more typical. The main reason for the term becoming shorter is the increased frequency of technology change and reductions in capital cost, therefore reducing the need to spread the cost over a long period of time.

ii. **The minimum volume threshold: the number of copies monthly, quarterly or annually that the customer commits to pay for**

A minimum print volume is used to calculate the base costs of the service and maintenance of the device based on the agreed monthly usage for that particular device. In calculating this number customers need to know their average volume capacity per month. While the number may be averaged over a longer period to manage peaks and troughs, it is important to be as accurate as possible so that the supplier can recommend the most appropriate device.

On this point, the word average is very important. We know print volumes change over any given hour, day, week or month so averaging should take into account the workloads and user traffic both currently and into the future. It can be difficult for many customers and sales organisations to predict future volumes. However, there is a process to gauge, manage and average out the print volumes to ensure more effective utilisation rates can be achieved.

The minimum monthly volume is the print commitment that you, the customer, will agree to pay for each month for the term of the contract. If you produce less volume in any given month you will still be charged for the minimum monthly print volume that you have committed to. Sometimes a print contract may specify quarterly or annual print volume reconciliations, which can be more effective in smoothing out monthly fluctuations by averaging volumes over a longer period.

iii. **The service and maintenance rate for the device or devices**

Once a device or devices have been selected by the sales company (matching off with the appropriate speed, features, functional capabilities and volume capacity required by the client) they use either their standard price book costs or calculate a bespoke service and maintenance contract.

The service costs can be calculated to include or exclude toner supplies, warranties, onsite service and parts, depending on the solution fit and negotiation. For this exercise, assume everything is included. The service and maintenance contract is usually given as a click charge, or fee, per click for black and white and a different, usually higher, click charge for colour.

Remembering this component of the contract is for servicing the device and providing all the toner they can eat at the given service rate. Say the service click rate charge is $0.008 for black and white.

Multiply that by the number of pages printed each month.

All three ingredients are calculated to provide a single page per print rate (i.e. B&W and/or colour). This amortised rate will always be higher than the base service and maintenance rate charge to support the devices in-field. The contract will operate over the agreed term and is offset against the agreed monthly minimum volume commitment.

3. Variations and how they can impact the customer

Variations to an existing contract may be easy to describe and sell but the impact of variations going wrong will has enormous ramifications and costs to the unsuspecting customer. The anchor that becomes attached to the business through adding devices, increasing minimum print volumes or raising the page per print charge can be crippling to an organisation.

Sales organisations that are good at selling a page per print contract are usually the most effective at adding variations to a contract. This practice has both a cause and effect that the end customer needs to fully understand.

A variation can occur by adding additional devices to a contract either as part of an existing master contract or as an additional, linked, contract for the same customer. There are three ways a variation can be changed.

First is that the agreed minimum print volumes can be increased, depending on the product sold. The change may be to the total contracted volume or the device minimum.

The second is that the total page per print contract can increase. As the increase is usually only a marginal amount each time a device is added, the client never sees the real impact that this makes to the total contract price.

The third option is to increase the length of the current con-tract term. In these situations the increase can extend for all the existing devices that are currently under contract. This achieves two things: it reduces the need to increase the page per print charges and it gives a windfall to the sales or finance company by increasing the sale price of the overall contract by extending the entire contract

The sales company has the ability to use each of the three ingredients individually or in combination.

Other cause and effect situations exist but these are the most important to understand.

4. The good and bad of a PPP contract

PPP contracts are not all bad news but you need to know when best to use them, when to recommend them and, most impor-tantly, when not to. Unfortunately the page per print contract is one of the most profitable for the print industry so it is rare that they are not recommended.

A page per print contract is best implemented when a cus-tomer believes they will have little change in their usage patterns. This keeps print charges consistent and a known cost to the business. Also, due to its simple charging model, it allows easy straight-through charging to internal customers in the business.

What works against the page per print contract is that reduc-tions in the market service rate during the contract period are difficult for the customer to recoup. As the customer has locked themselves into a fixed page per print contract, even if the cost of the service and maintenance reduces in the contract period, the customer will find it difficult to receive a pricing reduction as the finance contract has already been set. Alternatively this can be managed through non-standard contract terms – such as those with staggered end dates – but if you have combined all contracts into a single end term and it has already been extended, you may be substantially disadvantaged.

My biggest gripe with a page per print contract is when a client is an acquisition target or plans to downsize or (in today's speak) right size. When these organisations commit to minimum print volumes their ability to divest, or downsize, parts of the business can be heavily impacted by the print commitment, which remains, regardless of how much they actually print. Some sales organisations will try to assist businesses in this situation but, as the arrangement is actually with the finance company, the sales company may be limited in what it can or is prepared to do.

5. What makes you attractive?

What makes you attractive is very simple indeed. Let me give you a hint. It's where the industry really makes its income.

It's all about the toner.

What makes you attractive, as a customer, is how much volume you are doing. Do your homework and understand your black and white and colour monthly print volumes. Colour is approximately 10 times more expensive than black and white so having the right statistics is a great starting point. The more volume you produce, the better you can negotiate. Device units are important but volumes count more over the term of the service contract.

NOTES

1. HP Press Release, A Finnie, "TFTNB: The First DeskJet Printer", 27 August 2010, http://h20435.www2.hp.com/t5/The-Next-Bench-Blog/TFTNB-The-First-DeskJet-Printer/ba-p/58076#.UOnXi8bXFCB

2. HP Press Release, "HP Celebrates Shipment of 200 Millionth HP LaserJet Printer", 12 November 2013, http://www8.hp.com/us/en/hp-news/press-release.html?id=1526950#.UOnbQsbXFCB

3. Darwin Correspondence Project, http://www.darwinproject.ac.uk/six-things-darwin-never-said

4. Stuart Jacobs, Regional IT Manager of Sims Metal Management, Customer, conversation with author, 15 October 2013

5. IDC Press Release, "Worldwide Page Volume Continued Slow Decline in 2012 as Gains in Developing Regions Failed to Offset Slowdown in Developed Regions, According to IDC", 29 July 2013, http://www.idc.com/getdoc.jsp?containerId=prUS24240913

6. iTWire "Booming digital economy boosts Australia Post" Graeme Philipson, 21 October 2013, http://www.itwire.com/it-industry-news/market/61962-booming-digital-economy-boosts-australia-post

7. White Paper by Francois Ragnet: The "Less Paper" Office: How to Reduce Costs, Enhance Security and be a Better Global Citizen. http://www.xerox.com/downloads/usa/en/t/TL_whitepaper_less_paper_office_Francois_Ragnet.pdf

8. Photizo Group Releases, 2013 Imaging Hardware Forecast, 29 July 2013, http://photizogroup.com/photizo/photizo-news/press-releases/2013-imaging-hardware-forecast

9. IDC Press Release, "Worldwide Page Volume Continued Slow Decline in 2012 as Gains in Developing Regions Failed to Offset Slowdown in Developed Regions, According to IDC", 29 July 2013, http://www.idc.com/getdoc.jsp?containerId=prUS24240913

10. Digital Distractions In The Classroom: Student Classroom Use Of Digital Devices For Non-Class Related Purposes, Bernard R. McCoy, University of Nebraska-Lincoln, Journal of Media Education, Vol. 4 - Number 4 page 5, October 2013, http://en.calameo.com/read/000091789af53ca4e647f The Reading Brain in the Digital Age: The Science of Paper versus Screens , Ferris Jabr, Scientific American, Vol. 309 – Issue 5, 11 April 2013, http://www.scientificamerican.com/article/reading-paper-screens/

11. IDC Press Release, "Worldwide Page Volume Continued Slow Decline in 2012 as Gains in Developing Regions Failed to Offset Slowdown in Developed Regions, According to IDC", 29 July 2013 , http://www.idc.com/getdoc.jsp?containerId=prUS24240913

12. Gartner Press Release, "Gartner Says Worldwide Traditional PC, Tablet, Ultramobile and Mobile Phone Shipments On Pace to Grow 7.6 Percent in 2014", 7 January 2014, http://www.gartner.com/newsroom/id/2645115

13. Photizo Group Release, "2013 Imaging Supplies Forecast 360, September 2013, By the Numbers" http://photizogroup.com/wp-content/uploads/2014/03/by-the-numbers-flyer.pdf

14. Recharger Magazine, "If you build affordable colour printing, will they come", 1 October 2013

15. Photizo Group Release: "Printer Market Continues to Contract as the Need to Print Lessens; Manufacturers that Adapt Will Compete Effectively", Larry Jamieson, 30 July 2013, https://photizogroup.com/?s=printer+market+continues

16. My Print Resource, "2012-2017 Worldwide MPS Forecast Still Shows Excellent Growth in Specific Geos and Segments"

17. Randy Dazo, 1 July 2013, http://www.myprintresource.com/blog/10979327/2012-2017-worldwide-managed-print-services-forecast TechNavio Report, Global Managed Print Services (MPS) 2012 – 2016, 4 October 2013, http://www.technavio.com/report/global-managed-print-services-mps-market-2012-2016

18. IDC Press Release, "Asia Pacific (excluding Japan) Print Services Market to Cross US$6 Billion Barrier by 2017", 18 January 2014, https://au.finance.yahoo.com/news/asia-pacific-excluding-japan-print-130000894.html

19. Photizo Group, Forecast: 2013 Managed Print Services (Asia Pacific), Ken Stewart, 13 September 2013

20. Ibid.

21. ITwire Article, "Internet of Things – 26 billion by 2020", Graeme Philipson, 16 Dec 2013, http://www.itwire.com/business-it-news/technology/62636-internet-of-things---26-billion-by-2020

22. *Blue Ocean Strategy,* Professors W. Chan Kim and Renee Mauborgne, first published in 2005, Harvard Business Review Press, http://www.blueoceanstrategyaustralia.com.au

23. *Strategic Management Concepts and Cases* by AA Thompson and AJ Strickland, first published 1999, McGraw-Hill

24. *Blue Ocean Strategy*, ibid.

25. IDC Abstract: "Australia Business Process Outsourcing 2013–2017 Forecast and Analysis", August 2013, http://www.idc.com/getdoc.jsp?containerId=AU2577407V

ACKNOWLEDGEMENTS

My thanks go to the special people who have helped shape my views and supported me with their contributions to this book:

Phil Hurley, who brings great insight, care and consideration to any topic or question presented to him. Dr David Cooke, who always provides the right counsel with a measured and authentic approach that leaves you wanting more. Stuart Jacobs, who is always willing to share his practical experience of the industry and its challenges, giving his perspective as an end client as well as a manufacturer. John Hall, who brings a sense of calm and warmth when discussing both the challenges and future opportunities that Australian OEM operations are facing. Melinda Tippett, a passionate, intelligent woman with an enormous amount of energy and drive who is committed to achieving measurable results for her business and who always 'tells it like it is'. And David Richter, an experienced hand at the tiller who assumes nothing and always asks the right questions.

Special mention must go to the journeymen who have provided me with a clearer road map on my journey to writing this book.

They are Roger Amir, a true, lifelong friend who gave me my first real opportunity to be part of an exciting and intriguing industry. Trevor Lewis, an esteemed English aristocratic, who gave me the keys to the car and the confidence to set forth on a journey across the office printing industry. And Christophe Lambert, my French entrepreneurial friend and colleague who inspired and developed a higher level of creative thinking.

My mentor, Peter Williamson, who continually provides an enormous amount of practical advice, support and encouragement, and who is never too far away to push my cause. My German counterpart Mr Peter Strohkorb whose level-headedness keeps me grounded. Two great guys who know how to explain IT simply, Shane Stewart and Matthew Stanton. My American friend, Ed Crowley, a true industry crusader and thought leader who is ahead of his time. Thanks also to Ken Stewart, Greg Walters and my MPSA education team members headed by Jennifer Shutwell. Thanks for those regular early morning phone hook ups and healthy debates with team members including Kevin De Young, Steve Spencer, Dave Fabrisio, Aldo Spensieri to mention a few – all with great insight into the challenges facing the industry.

This book would simply not have been possible if I didn't have the direction and inspiration from a real powerhouse in the Australian publishing industry who, as a successful author himself, has provided the framework, guidelines and focus I needed to get the job of writing and publishing this book done. Thank you Andrew Griffiths, I owe you.

To a team who put the finishing touches on my aspiration, Dominique Antarakis and Damian Clarke from The Copy Collective, you just know how to deliver. Dominique, thank you for helping project manage what I needed to get done and Damian, thank you for making the experience a painless one.

Thank you to the entire team at OpenBook Creative, especially the imaginative and resourceful Julie Renouf who took me under her wing and allowed me to bring my creation to life.

Last but certainly not least a big thank you to all my industry colleagues in Australia and across the globe, and more importantly to all the customers for giving our industry the opportunity to serve you and hopefully contribute positively to your business. Without you, we would not have an industry.

ABOUT MITCHELL FILBY

Mitchell has spent 25 years working in both the IT and the print and imaging industries with major original equipment manufacturers (OEMs) such as Oce, Kodak, Fuji Xerox and Canon, as well as private and publicly-listed IT companies in Australia.

In these companies he has held various senior executive roles including GM and CEO roles across varying parts of the business including sales, marketing and operations as well as full P&L responsibility of the business themselves. In this capacity Mitchell has been in a position to see firsthand how the print and imaging industry has cycled through both good and bad times. He has seen the transition from B&W to colour coping, analogue to digital printing, transactional to solution to consultative sales and more recently from a product-led to services-led models.

Shortly after completing a Master of Business (MBA), majoring in Strategic Management at the University of Technology in Australia in 2006, Mitchell established First Rock Consulting, an independent consulting and advisory business for both large and enterprise-end clients.

Through First Rock Consulting Mitchell assists many customers' gain greater transparency of their actual office printing operations and the costs associated to this somewhat hidden or unknown element. Through effective analysis and industry insight First Rock Consulting provide a spotlight on the real costs of office printing and document processes.

More and more Mitchell is advising clients on the process of transitioning from the physical world of printed output to a more efficient digitised ready content. First Rock consulting aims to help businesses unlock the knowledge residing in legacy hard copy documents as well as improving business workflow and optimising processes to streamline and increase user and business productivity and performance.

As a renowned industry expert Mitchell also works with many of the leading OEM manufactures and managed print and IT service providers who either require a 3rd party independent or advice on how they best transition to a services-led model. In fact First Rock Consulting are considered the "Switzerland" of the industry. This is supported by Mitchell's active involvement in the Managed Print Services Association (MPSA – the peak independent body for the office printing and document imaging industry) where his takes his role on the education sub-committee as both a duty and responsibility to the industry.

On a personal note, Mitchell continues to have an enormous drive, energy and passion in everything he goes after. Although early in his adult life was shaped around extensively travelling the world, and with a long sporting background across many disciplines such as Surfing, Soccer, Rugby Union and Rugby League, Outrigging Canoeing, Surfboat rowing both in Australia, United States and in the United Kingdom. Today his real focus and passion is keeping up with his two children who just as passionate in their pursuits. As Mitchell continues to realise the apple truly does not fall far from the tree.

FIRST ROCK CONSULTING

Contact: www.first-rock.com
Email details: mitchell.filby@first-rock.com

OUR SERVICES:

First Rock Consulting specialise in the office printing and document imaging industry. As a truly independent authority or as some global industry analysts say, we are the "Switzerland" of the Australian industry as we work with both end clients as well as OEM's and Industry providers.

Our consulting services are shaped around helping clients better understand the changing technology requirements and the costs of supporting and servicing these operations. This encompasses a business's existing paper based output and processes to a more efficient digitised process and workflow so that information and data can be more effectively shared and utilised within the business.

We (First Rock Consulting) are experts in indentifying and analysing a business's office printing and document imaging environment. Our experience, knowledge and industry insight allows us to advise customers on industry benchmarking, best of class practices, cost savings initiatives, consolidation and asset utilisation, technology investment, change management, IT and vendor management and much more.

Many of our programs start by just analysing our customer's current expenditure on office printing. Many clients feel or know they are spending too much, but identifying and determining what those costs are and how to calculate those various elements is the challenge of time, executive appetite and the financial return on their effort.

To assist our clients we work with them to identify the "size of the prize" to really understand if the effort meets the reward.

Through this process we provide customer's greater transparency and insight to how they are being charged and how they can improve their business performance through a practically applied phased approached. In all cases clients realise that they are paying too much and engaged us to help either;

1. Reduce operational costs,
2. Improve technology deployment
3. Assist in changing user behaviour.
4. or all of the above

These three pillars are fundamentally core to First Rock Consulting transformational framework model. We identified that the three key elements of consistent and sustainable change had to deliver on and meet the business outcomes required around financial, technology and people as the key drivers for continued business performance.

To find out more please go to www.first-rock.com and simply complete your name and email address or email me directly on mitchell.filby@first-rock.com.

www.ingramcontent.com/pod-product-compliance
Lightning Source LLC
Chambersburg PA
CBHW060401220326
41598CB00023B/2990